Smith-Kettlewell Eye Research Institute

Visual Function and Its Management in mTBI

A national conference held at

**The Smith-Kettlewell Eye Research Institute
San Francisco**

March 4-5, 2011.

Edited by Christopher W. Tyler

Visual Function and Its Management in mTBI

Proceedings of national conference on the management of functional visual deficits in mild traumatic brain injury (mTBI), held at the Smith-Kettlewell Eye Research Institute, March 4-5, 2011.

Editor: Christopher W. Tyler, PhD, DSc

Casewrap hardcover: 172 pages. Acronym list, references, index.

ISBN: 978-0-9890819-0-0

Cover photo: Shubhangi Kene

Special thanks to Diane P. Remillard for her help in the preparation of this publication.

Smith-Kettlewell Institute

Visual Function and Its Management in mTBI

March 4th and 5th, 2011

Organizing Committee

Arthur Jampolsky

John Brabyn

William Good

Christopher Tyler

Glenn Cockerham

Gregory Goodrich

Ronald Schuchard

Bebe St. John

Program

Friday March 4

Welcome   Arthur Jampolsky, OD, MD

Introductory Remarks

Col Donald A. Gagliano, MD, MHA, CPE, FACHE
>   DoD Principal Advisor for Vision, Director
>   DoD/VA Vision Center of Excellence

Col (Ret) Robert A. Mazzoli, MD, FACS
>   Director, Education, Training, Simulation and Readiness
>   DoD-VA Vision Center of Excellence

Francis L. McVeigh, OD, FAAO, MS
>   Senior Clinical Consultant, IPA
>   Telemedicine & Advanced Technology Research Center (TATRC)

Glenn C. Cockerham, MD
>   Chief, Ophthalmology and Eye Services,
>   VA Palo Alto Health Care System

Session 1: The Nature of the Injury

Overview and orientation	William Good, MD
	Christopher Tyler, PhD, DSc
Impacts on vision function – including early and late onset - what we know and don't know	Gregory Goodrich, PhD
Blast-induced ocular and visual changes	Glenn Cockerham, MD
New information learned from experimental blast TBI	Matthew Harper, PhD
Impacts on hearing	Gabrielle Saunders, PhD

Panel Discussion
>   William Good, MD (Moderator)
>   Gregory Goodrich, PhD
>   Glenn Cockerham, MD
>   Matthew Harper, PhD
>   Gabrielle Saunders, PhD
>   Stephen Heinen, PhD

Saturday March 5

Session 2: Tests, Evaluation and Assessment

mTBI Diagnosis: Objective measures of visual dysfunction	Randy Kardon, MD, PhD
Light sensitivity in mTBI: Models for assessment and causality	Michael Gorin, MD
Self-assessments and their effectiveness: What vision tests are currently used?	Gregory Goodrich, PhD
Oculomotor function tests, photosensitivity, accommodation and convergence testing in mTBI	Suzanne Wickum, OD
Functional imaging for oculomotor deficits in mTBI	Christopher Tyler, PhD, DSc
Concussion diagnosis	Anne Mucha, PT, MS, NCS

Panel Discussion
> Arthur Jampolsky, OD, MD (Moderator)
> Randy Kardon, MD, PhD
> Michael Gorin, MD
> Gregory Goodrich, PhD
> Suzanne Wickum, OD
> Christopher Tyler, PhD, DSc
> Anne Mucha, PT, MS, NCS
> Yury Petrov, PhD

Session 3: Therapy

Current practices	Glenn Cockerham, MD
	Gregory Goodrich, PhD
Accommodative and convergence training	Kenneth Ciuffreda, OD, PhD
Medical or ophthalmic interventions in the globe	Kim Cockerham, MD
Comparison with management of hearing impacts	Gabrielle Saunders, PhD

Panel Discussion

John Brabyn, PhD (Moderator)
Glenn Cockerham, MD
Kim Cockerham, MD
Kenneth Ciuffreda, OD, PhD
Gabrielle Saunders, PhD
Pia Hoenig, OD

Future Research Funding Possibilities in mTBI and Vision Function
James Jorkasky, Executive Director, National Alliance for Eye and Vision
Research (NAEVR)

Remaining Problems, Unmet Needs, Gaps and Funding Sources
(Discussion by DoD, VA and other participants)

Wrap-up Session
(Closing comments by the organizers and others)

Attendees

Dan Adams, PhD
Assistant Professor of
Ophthalmology
UC, San Francisco
AdamsD@vision.ucsf.edu

Felix Barker, OD
DoD-VA Vision Center of
Excellence
Felix.Barker@va.gov

John A. Brabyn, PhD
Executive Director
The Smith-Kettlewell Eye Research
Institute
brabyn@ski.org

Kenneth J. Ciuffreda, OD, PhD
Distinguished Teaching Professor
SUNY/State College of Optometry
kciuffreda@sunyopt.edu

Glenn C. Cockerham, MD
Chief, Ophthalmology and Eye
Services
VA Palo Alto Health Care System
cockerham@stanford.edu

Kimberly Cockerham, MD, FACS
Associate Clinical Professor
Stanford University Department of
Ophthalmology
VA Palo Alto Health Care System
cockerhammd@gmail.com

James M. Coughlan, PhD
Scientist
The Smith-Kettlewell Institute
coughlan@ski.org

Bill Crandall, PhD
Scientist
The Smith-Kettlewell Institute
bc@ski.org

Sony Devis
Controller
The Smith-Kettlewell Institute
sony@ski.org

Anas Elsaid, MD
Research Associate
The Smith-Kettlewell Institute
Elsaid_anas@yahoo.com

Col. Donald Gagliano, MD
VCE Director
Vision Center of Excellence
Donald.Gagliano@tma.osd.mil

Debby Gilden, PhD
Senior Scientist
The Smith-Kettlewell Institute
debby@ski.org

William V. Good, MD
Senior Scientist
The Smith-Kettlewell Institute
good@ski..org

Gregory Goodrich, MD, PhD
Supervisory Research Psychologist
VA Palo Alto Health Care System
Gregory.Goodrich@va.gov

Michael B. Gorin, MD, PhD
Harold & Pauline Price Professor
of Ophthalmology
Jules Stein Eye Institute
gorin@jsei.ucla.edu

5

Matthew Harper, PhD
Research Scientist – Neurobiology
Iowa City VA Center for Prevention
& Treatment of Visual Loss
mharper@iastate.edu

Stephen J. Heinen, PhD
Senior Scientist
The Smith-Kettlewell Institute
heinen@ski.org

Pia Hoenig, OD, MA, DBO, FAAO
Chief, Binocular Vision Clinic
University of California, Berkeley
mphoenig@berkeley.edu

Arthur Jampolsky, MD
Founder
The Smith-Kettlewell Institute
aj@ski.org

James Jorkasky
Executive Director
National Alliance for Eye and
Vision Research
jamesj@eyeresearch.org

Randy Kardon, MD, PhD
Professor and Director of Neuro-
ophthalmology
Iowa City VA Center for Prevention
& Treatment of Visual Loss
randy-kardon@uiowa.edu

Anthony Kontos, PhD
Associate Professor, Department
of Orthopedic Surgery
UPMC/U of Pittsburgh Schools of
the Health Sciences
akontos@pitt.edu

Mihir Kothari, MD
Visiting Scholar
The Smith-Kettlewell Institute
drmihirkothari@gmail.com

Lora T. Likova, PhD
Associate Scientist
The Smith-Kettlewell Institute
lora@ski.org

Lori A. Lott, PhD
Research Associate
The Smith-Kettlewell Institute
lott@ski.org

Diana P. Ludlam, BS, COVT
SUNY/Optometry
dianaeye@aol.com

Manfred MacKeben, PhD
Scientist
The Smith-Kettlewell Institute
mm@ski.org

Col. (ret.) Robert Mazzoli, MD, FACS
Director, Education, Training, and
Simulation
DoD-VA Vision Center of
Excellence
robert.mazzoli@us.army.mil

Francis L. McVeigh, OD, FAAO, MS
Senior Clinical Consultant, IPA
Telemedicine and Advanced
Technology Research Center
francis.mcveigh@tatrc.org

Ming Mei, PhD
Postdoctoral Fellow
The Smith-Kettlewell Institute
ming@ski.org

Anne Mucha, PT, DPT, MS, NCS
Assistant Director Neuro/
Vestibular Services
University of Pittsburg Medical
Center
muchaa@upmc.edu

Spero Nicholas, MS
Programmer Analyst
The Smith-Kettlewell Institute
spero@ski.org

J. Vernon Odom, PhD
Professor of Ophthalmology
West Virginia University
Eye Institute
odomj@wvuhealthcare.com

Yury Petrov, PhD
Assistant Professor
Northeastern University
Y.Petrov@neu.edu

Pieter Poolman, PhD
Iowa City VA Center for Prevention
& Treatment of Visual Loss
pieter-poolman@uiowa.edu

Laura Renninger, PhD
Associate Scientist
The Smith-Kettlewell Institute
laura@ski.org

Tina Rutar, MD
Assistant Professor of
Ophthalmology
UC, San Francisco
RutarT@vision.ucsf.edu

Gabrielle Saunders, PhD
Investigator & NCRAR Deputy
Director for Education
National Center for Rehabilitative
Auditory Research
Gabrielle.Saunders@va.gov

Marilyn E. Schneck, PhD
Scientist
The Smith-Kettlewell Institute
mes@ski.org

Ronald A. Schuchard, PhD
Research Career Scientist
VA Palo Alto Rehabilitation R & D
Service
rschuch@stanford.edu

Bebe St. John
Research Administrator
The Smith-Kettlewell Institute
bebe@ski.org

Ender Tekin, PhD
Research Associate
The Smith-Kettlewell Institute
ender@ski.org

Jeffrey Tsai, MD, PhD
Associate Scientist
The Smith-Kettlewell Institute
jeff@ski.org

Christopher W. Tyler, PhD, DSc
Senior Scientist
The Smith-Kettlewell Institute
cwt@ski.org

Suzanne Wickum, OD, FAAO
Clinical Associate Professor
University of Houston
SWickum@Optometry.uh.edu

Session 1: The Nature of the Injury

Overview and Orientation
William V. Good, MD and Christopher W. Tyler, PhD

Impacts on vision function – including early and late onset—what we know and don't know
Gregory Goodrich, PhD

Blast-induced ocular and visual changes
Glenn C. Cockerham, MD

New Information Learned from Experimental Blast TBI
Matthew Harper, PhD

Impacts on Hearing
Gabrielle Saunders, PhD

Overview and Orientation

WILLIAM V. GOOD, MD and CHRISTOPHER W. TYLER, PhD

>> WILLIAM GOOD:

I would like to an add my welcome to everyone here and particularly my thanks to all of you for taking the time out for what I know are very busy schedules to attend this meeting. We are really pleased by the depth and quality of the people we have been able to attract to come to the conference. I know there will be interesting interchange to go with the panel discussions and brief lectures.

It seems that there isn't a day that goes by that I do not open up the newspaper and read something about a traumatic brain injury. It is in many of the sports: Hockey injuries. Baseball. Football injuries. And, of course, the military injuries, which are really paramount.

Even in my regular pediatric practice I see children who have concussions at least one or two times a week. They come to the office and some have symptoms I do not understand. Given this, I am really pleased to be on the education side of the conference and hope to learn a lot from all of you.

Again, let me extend my thanks to all of you for attending. With this, let me turn the podium over to Chris Tyler who has a few introductory comments as well.

>> CHRISTOPHER TYLER:

In my presentation I wanted to do something to help us think about the kind of mechanisms that are involved in TBI injury. I am going to analyze one viewpoint about what the mechanisms might be, that focuses on the deep brain structures, the sites of the core brain injuries in mTBI.

My inspiration came from some computer modeling of head impacts in football injuries. These are helmeted heads impacting and being videoed from several viewpoints, which enables the reconstruction of the impact both externally and within the brain structures. The outcome of this 2007 study by Viano et al. shows the physical impact within the brain according to whether the injury was non-concussive or concussive.

When no concussion ensued from the impact, the distribution of sheer stresses inside the brain was rather uniform and small. It was not very strong. But, when there was a concussion, there is a very clear pattern of stresses that focus on the core of the brain: the corpus callosum and the brainstem region. This is a very telling analysis that then can be compared with MRI morphology study of the shrinkage or expansion of brain tissue in mTBI cases.

An average map for a number of mTBI cases may be derived by projecting them onto the standard MNI brain. The region of shrinkage is very much in the brainstem and central core regions of the brain and very little around the cortical regions.

This analysis may be from several viewpoints. If you view it in a coronal section the shrinkage is again focused in the core regions and the brainstem. You can see it again from an axial viewpoint; which gives you three dimensions of analysis. It really focuses our understanding of the damage very much both the physical forces and the tissue damage, giving very similar patterns of activation. This is the one single message I wanted to cover.

Let me emphasize that these are not functional data; these are structural imaging data. That is my whole story.

Discussion

>> AUDIENCE MEMBER:

Those were clinical patients?

>> CHRISTOPHER TYLER:

Yes, adult patients (in the second study).

>> AUDIENCE MEMBER:

What kind of function or dysfunction did they have individually or mentally? With those injuries?

>> CHRISTOPHER TYLER:

I do not know the cognitive diagnostics they applied.

>> AUDIENCE MEMBER:

They were all mild TBI?

>> CHRISTOPHER TYLER:

They were long-term, 8 months to two years post-trauma mTBI patients.

>> WILLIAM GOOD:

Now, it is my pleasure to provide introductions for our speakers this morning. I cannot possibly do justice with brief introductions to all of the accomplishments of the people who are speaking today.

First I will introduce Greg Goodrich, whom I have gotten to know over the last six months through organizational efforts to get the symposium together. Greg received his PhD in experimental psychology in 1974 from Washington State and had a career with the U.S. Veterans Administration. His current position title is Supervisory Research Psychologist at the Psychology Service and Western Blind Rehabilitation Center. He has published articles too numerous to count and is one of the experts in his field.

Impacts on Vision Function – Including Early and Late Onset: What We Know and Don't Know

GREGORY GOODRICH, PhD

I would like to also convey my thanks to John and Smith-Kettlewell for putting this event together. I would also like to thank all the previous speakers for saying all the things I wanted to say. I would like to tell you a little bit about the VA system of care.

VA Services

The VA is composed of 21 Veteran Integrated Service Networks (VISN) operating 153 medical centers. I won't detail specific services here, except to note that there are the 4 Polytrauma Rehabilitation Centers (PRC) that Glenn mentioned with an additional one planned to open in the near future. PRCs are in patient programs serving some of the most seriously injured troops as well as veterans with severe injury usually including a traumatic brain injury. Each VISN has a designated Polytrauma Network Site (PNS) that serves patients on an out-patient basis for general pain, general medicine, neuropsychology, audiology and vision, as well as referral for additional care. PNS patients typically diagnosed with mild TBI and/or post-traumatic stress disorder (PTSD), in addition to other diagnoses. The studies I will report on utilize patients from the Palo Alto Health Care System which houses both a PRC and PNS and which is unique in that it also has a Blind Rehabilitation Center serving blind and low-vision troops and veterans.

This means that one of the problems is for us to try to take the numbers that comes out of VA system to characterize troops injured in Afghanistan or Iraq. If you take numbers that are sliced across the four PRCs, you get one picture that typically includes major physical injury and TBI ranging from mild to severe. If you take a slice of those through the PNS clinics, where they are seeing mild traumatic brain injury, you get another picture. If you go down to the 153 medical centers, where there are relatively young, active duty troops as well as older veterans, you can get an entirely different picture of incidence and prevalence of vision loss and binocular/oculomotor dysfunction.

Diagnosis of mTBI

Last year, I did a SurveyMonkey questionnaire to some 600 VA optometrists. The response rate was about 33%, which I thought was quite good. Of those that responded, 76% were seeing patients with mTBI diagnoses. Some of them were doing routine optometric examinations and some were doing a more formalized binocular vision examination. Currently there is no defined binocular vision examination guidance for these optometrists.

So, to try to get through this slice of information in a little bit of a different way, in our studies that have looked at the Polytrauma Network site level we are finding about 80% have one or more vision problems. An unpublished paper I have just read from a polytrauma support center is finding rates that are a little bit lower than what we have reported in Palo Alto and those reported by Joan Stelmack at the VA Heinz Hospital near Chicago. It is still a significant number.

One of the things I think we do know enough about is the binocular dysfunctions in mTBI. They seem to be associated in that 80% of mTBI patients, by some reports, have one or more diagnosed binocular vision findings. Now, I have to get on my own soapbox and say, we do not know a lot about what an mTBI is. We have a procedure in place for diagnosing it, but this primarily revolves around self-reporting. If you have been in a blast that causes mTBI, we do not know until you actually report symptoms. My point is that because you are asking the person involved to self-report it is difficult to know how accurate the diagnosis is.

This afternoon I will talk more about some of the functional vision results that we have gathered through asking people about their vision. What strikes me is that, if you ask people coming in to a polytrauma network site clinic, "How is your vision," they generally will say it is fine. We do the visual acuity test and find that 97% will have near normal or normal visual acuity in a PNS. In a study of 161 consecutive patients we found 3% had less than normal visual acuity; some were legally blind, but the vast majority say they do not have problems with vision. The gold standard test, visual acuity, shows their visual acuity to be "normal." What could possibly be wrong? It is also interesting that if you ask them about visual symptoms such as reading difficulty, light sensitivity, and so on, a very large percentage will endorse these symptoms. Yet they don't attribute the symptoms to a problem with their eyesight.

We know the literature tells us that binocular dysfunction/ocular motor deficits are commonly associated with mTBI diagnosis. We will skip over the fact that, as one of my British professors would say, we use an "airy-fairy" definition of mTBI.

Still, studies at the VA suggest there is a high rate of binocular/oculomotor dysfunction related to mTBI. Over 80% of the people we are seeing in the PNS center are there with a diagnosis of mTBI related to one or more blast events.

I also have to caution you that the problem with self-reports is that there are a lot of unknowns. We can be sure, I think, that the individual was exposed to a blast event. Many other variables are harder to know: How big was the blast? How far away were they from the blast? Were they indoors, outdoors, in a vehicle, not in a vehicle? As Glenn alluded to, we also need to ask how does a blast injure the brain? Is that similar or different than other mechanisms of injury? In the long run this is an important question.

When we look at binocular and oculomotor dysfunction, the most common dysfunctions we find are accommodative dysfunction, convergence insufficiency, and saccadic dysfunction. We know these are important for a lot of everyday tasks: Reading, using computers, watching television. In severe cases you can have problems with balance. These are very, very critical sorts of things in terms of how well the individual's quality of life.

One of the anecdotal things I would like to tell you is that, while people have these dysfunctions, their perception of their vision is that it's good. And even though you may diagnose a binocular vision problem they may not come back to you until they get into a situation where reading becomes a problem for them.

Some of the people we see come in because they are now going back to school. They find that they are having trouble keeping up with the reading demands. Some will report that after five minutes they feel fatigued, are unable to keep their place, lines of text become blurry, etc. We also have people who get into work situations where they have trouble doing the visual aspects of the job.

In other words, they may feel fine initially after a deployment. Later on, however, the symptoms will start to emerge.

The other thing about this population is that they have an incredibly high rate of co-diagnosis. Especially among the mTBI population, post-traumatic stress disorder (PTSD) is common as are depression, drug or alcohol abuse, and other psychological symptoms

We are finding that there are probably some overlapping visual symptoms between PTSD and the binocular dysfunctions. The person may also have depression. If you have sleep problems you get into a chicken-and-egg sort of effect. Earlier it was mentioned we do not know a lot about the effects of different pharmacological agents routinely prescribed. For example, a group of people coming in for pain, depression, etc, etc is getting drugs. We are trying to study vision and we do not know these interactions so it can be problematic to say that light sensitivity, for example, is due to a visual dysfunction or due to the PTSD, or both.

I think the way to summarize my talk is to say that I am going to be talking here about mostly what I do not know. Our studies pretty consistently show that about 80% of PNS patients with mTBI have binocular oculomotor dysfunction, but we do not have control populations for this. On patient self-report we know a history of binocular dysfunction is pretty low, and that the rates we are seeing are higher than those reported in the civilian literature, but we need better research controls. And the patients we are seeing are not randomized. So we are taking a small subset of all the people deployed to Iraq and Afghanistan, where what are needed are studies that will allow us to extrapolate to the larger problem.

Symptomatology

Regarding symptomatology, as I mentioned, reading deficits are common. People will report reading problems and the majority of those who report note that this is new since their injury. Other symptoms they report are blurred vision at near and distance, floaters, light sensitivity, and an inability to do sustained reading, among others. They also note a difficulty remembering what was read and that their eyes tire while reading. But, again, the reporters typically do not associate these symptoms with their vision. It becomes a challenge for the researcher, the clinician, and others charged with caring for these patients. How do you identify the source of the symptoms; are they vision related or related to PTSD, prescribed medications, or other causes or combination of causes? It is not like I stubbed my toe and I show up to get medical attention. I am having problems and I am not sure how I am addressing them or how to go about addressing them.

So what don't we know? Well, and this again is going to be a bit repetitive, but we really don't know the prevalence and incidence. I think it is terribly important that we understand what we do not know because the unknowns help us to see the impact of the problem and to understand how to focus on resolving these problems. I think it is also important because the events associated with mTBI are not just limited to combat troops. All we have to do is look at San Mateo down the road from here, where there was a major natural gas pipeline explosion. Mining accidents, terrorist attacks in New York, London, Madrid, Bali, and car injuries may all lead to TBI and resulting visual loss and dysfunction. A myriad of things cause mTBI. I think we need to look at this comprehensively. We don't know a lot about the natural time-course of these injuries because to a large degree vision loss and dysfunction has not been a focus of those studying and treating brain injury and brain injury has not been a major focus of those studying vision loss and dysfunction.

We are pretty sure that most people who have an mTBI with an immediate vision problem seem to spontaneously resolve it. But we do not know how long that spontaneous resolution takes. Is it a day? Is it a week? Is it months? If it will spontaneously resolve in three months, you do not want to spend a lot time treating that person's vision symptoms during that initial three month period because it may be just wasting your time and their time and resources.

We also do not know whether — and this is speculation — you can have an mTBI and not have immediate visual symptoms but then have the symptoms manifest later. If this occurs it would change current practice regarding vision and mTBI. We do not even have a gold standard assessment and examination that is widely used, so that when we start comparing numbers from different studies we can't be sure if we are comparing apples and apples or apples and oranges. In fact, even within an individual VA, different assessment standards are used. Some of this is just differences between two clinicians. But, there also can be a difference between a visit to an eye clinic as compared to the low vision clinic, where the patient may receive very different examinations. So, all of the incidence rates that we're talking about at this symposium must be seen in light of these uncertainties. We'd like to say that we have a gold standard comparison, but we are lacking that gold standard and developing it should be a priority.

Future Goals

There is a good deal that needs to be done. We do not fully know the impact of binocular dysfunction on an individual's life, although the available evidence suggests it can be substantial. Do people change their lives because of their symptomatology? I have difficulty reading and eye fatigue. Do I stop reading? What is the affect of that decision over the course of an individual's life? Does it decrease earnings potential; does it affect social relationships; does it lead to other psychological problems?

This reminds me of another point that I want to make. The mTBIs that we are primarily focusing on in the DoD and the VA are happening to people between 19 and 36 years of age. Whatever are the consequences of these events, they are going to last over the course of that person's lifetime. This is potentially a lifelong problem unless we can intervene in some way to remediate or eliminate the problems. We do not really know how much the problem impacts their daily life. Nor are we really sure what the best treatment or rehabilitation strategies are.

Talking out of turn about the VA here, I think that VA clinicians and therapists do the best they can for the individual patient, but there is not a lot to guide them in that effort. Therefore, we do not know how successful their efforts are. In fact, we may not know enough to know how to measure the success. Is a resolution of a clinical symptom a success? We have many unknowns and I believe that symposium such as this one sponsored by Smith-Kettlewell provides an excellent forum to discuss them and attendees who are well equipped to develop answers.

There are the couple of things we do know. One is that the troops that are coming back represent a really unique opportunity to advance our knowledge base and our understanding of vision and mTBI. We also know that we need to reduce the unknowns in order to meet with the DoD and VA mandates to provide care to our wounded warriors. More generally, we also have an opportunity to really advance things in this area and to do so in a way that will affect the civilian world as well. This is the history of military medicine. Military medicine responds to instant demands and needs to produce great solutions. These then trickle out into the broader community, so everyone can benefit from it. These are my thoughts on the many things I do not know sprinkled with a bit of what I think I know.

If you have any questions that you would like me to answer about either what I may know or what I don't know, I would be happy to give it a try.

Discussion

>> WILLIAM GOOD:

Glenn, why don't we have you come set up while we take some questions. We should have subtitled our conference "what we don't know" or something like that.

>> DONALD GAGLIANO:

That is a great point, because we do have another conference that we call the Military Vision Research Symposium. It is looped to the Schepens meeting, which is all about the gap identification and through the TATRC symposia. It is the framework for the announcements that go out with the requests for proposals. What we are doing here is actually deriving the foundation for the requests for proposals that you may be seeing in future announcements for Department of Defense funding. So it is important for those of you in the research fields that there is full communication. That helps us help you.

>> AUDIENCE MEMBER:

I just want to make one remark regarding something that Bill said: people having changes in their activities without recognizing that it matters to women with breast cancer and looking at a variety of things for drug toxicity. We found a very good correlation between how many activities they stopped doing, but did not attribute to their vision problems or align with their complaints about their vision. In other words, there were a whole separate set of lifestyle changes that people were making without the awareness that they were doing so because of vision related events. In fact, they denied this happened. I think we are often looking at the tip of an iceberg when we ask patients to tell us how vision is affecting their lifestyle.

>> WILLIAM GOOD:

Thank you. That is an excellent point.

It is a pleasure to now introduce Glenn Cockerham. He is the Chief of Ophthalmology and Eye Services at the VA hospital in Palo Alto center. He is also a Clinical Associate Professor of Ophthalmology and Pathology at Stanford and a former colonel in the United States Air Force. He is going to add more to what we know and what we do not know in his talk this morning titled "Blast-induced ocular vision changes."

18

Blast-Induced Ocular and Visual Changes

GLENN COCKERHAM, MD

The Spectrum of Blast Injuries

I will try to tell you a little bit of what we do know. As mentioned before, at Palo Alto we have one of the four polytrauma centers that was set up to start receiving patients in 2005. In 2006 we were approved by the Institutional Review Board (IRB) to do a prospective study that I am going to talk about here.

Our center has the unique opportunity to look at a group of military members several weeks or months after their initial injury. It is possible to bring them into the clinic and do a thorough evaluation. Whether they were in Walter Reed or Bethesda, or in a drug induced coma, if they survived and go to the polytrauma centers, we have a period of months to work with them, sometimes shorter.

Tomorrow we're going to talk about the methodology of what we are doing. We standardize room lighting, who examines the patient, what we do (our examination includes neuropathy, glaucoma), and we have dedicated support staff as well.

We do not know the overall denominator. I do not know whether our research sample is representative or not. There are reasons that I think it may be representative. Our hypotheses are very similar. We believe that a blast force sufficient to cause a TBI through stress, shear waves, acceleration, deceleration, or whatever the effect is, causes intraocular damage, which may be open globe, a penetration from metal going through the eye, for example.

The first responders pick up these injuries. These open globes are closed primarily in Afghanistan or Iraq. There is a tragedy with all of the open globes in the current conflicts. There have been a couple of thousand reported, versus a couple of hundred thousand who have been exposed to blasts. There is a huge gap between the eye injuries we know about and those we do not know about. In other words, an injury to the eye is only detected by a thorough examination because you might not pick it up on the battlefield. Secondly, a blast that causes TBI may cause visual dysfunctions on its own, or the blast can cause damage to the optic nerves,

19

the tissues, the eyelids, or the adnexa and cause defects through that. These are very simple hypotheses.

There are several forces embedded in a blast wave. The primary or shock wave is huge if you're close enough to it. It decreases logarithmically as you get away from the center of the blast. It is supersonic and embedded within are the high and low frequency waves that can fracture bone and cause cavitation at tissue interfaces. These are particularly troubling, within the vitreous, the intestines, or a blast lung. It can rupture intestines or eardrums. Any air filled organ is susceptible.

Now the brain is different, being encased in bone and cerebrospinal fluid. And with the right protective armor, polycarbonate will stop low velocity projectiles. Certainly penetrating eye injuries have decreased since it became mandatory to wear eyeware off the shelf. This became mandatory in Iraq in 2005. At that point in time, commanders were told that they would be court-martialed if the troops did not wear them, so they wear them.

The question is, does this gear work for blast waves? We'll look at that. Also, there are other effects. These blasts are pretty significant. I will show you a blast that was buried in an asphalt road. You can see what it can do. That is a huge effect on asphalt. Think about that. You can see the big rocks that were thrown up. Behind them, we see a forensics teams sent out to evaluate the blast characteristics.

The next blast was not in the lethal zone for all, although here were a couple people killed by this blast. Here is a forensic team trying to figure out how this blast happened. Here is road debris that was thrown 50-100 meters away. You can imagine how severe this concussive force is. It is enough to blast the treads off a tank.

Prospective Observational Cohort

We approached everybody who met our inclusion criteria and asked them to be in the study. A major inclusion criterion was a diagnosis of any level of TBI as assigned by DVBIC. There is a project manager at VA Palo Alto, like all VA facilities, that tracks the people with TBI. They assign the severity level for TBI. As a third party, we are not involved in assigning the TBI grade. We do not influence this in any way. All we do is approach them. They can enter the study or not. The other criterion is that they are able to come to an examination. We can not do a complete examination on a bedbound patient. They have to be able to go to the eye clinic and go through the tests.

We do ocular examinations in teams. There are several vision function tests that we do, and also the quality of life questionnaires. We exclude open globe injury questions from our studies. There are nine patients with the open globe injury in the fellow eye in the study. They are able to be in the study if they have a fellow eye that was exposed to blast and did not have an open eye injury.

This is an ongoing study. It is a young population, mainly men and mainly from operation Iraqi Freedom. We only had four from Afghanistan, which was called Operation Enduring Freedom at that time. The median time from injury in theater to the time that we did our baseline exam was as long as four months during the Iraq operation. More recently, we are getting referrals in as little as 3-4 weeks.

Everything has changed now with Afghanistan and the Surge, and Walter Reed and Bethesda National Naval Hospital are transferring patients earlier. The population keeps changing. We initially had relatively equal numbers of mild TBI, moderate-severe TBI, and penetrating TBI. Most were using eye armor. Some were not and some were unconscious and don't remember. There is also 'mounted' versus 'dismounted'. Mounted means that are hit while in a vehicle. 'Dismounted' means you are on foot, like the guy that was in the roadbed. His patrol was on foot. It also could have been a vehicle that was blown over by that blast.

Ocular Injuries

Now I am going to talk about ocular injuries, particularly closed-globe injuries. We define closed globe injuries by anatomic injury zone. Zone 1 is conjunctiva and cornea. Zone 1 is external. Zone 2 is trabecular meshwork, iris and lens. Zone 3 is vitreous, choroid, retina and optic nerve.

The trauma eye center in Birmingham has defined our terminology for closed globe injuries and the three zones are based on their definitions. Forty-three percent of the patients out of that initial group of 46 had internal or external injuries, external being Zone 1. These injuries were pretty well spread across all three zones. Some had multiple zone injuries. Highly correlated with closed-globe injuries were facial fractures, and facial or eyelid scars.

What was not correlated was the severity level of TBI. Why do we care about mild TBI (mTBI)? You know that in mTBI that there are no other injuries. Oh really? Please tell me where that reference is. We do approach statistical significance for closed-globe injury in mTBI, with a p value of 0.07. We found no correlation between eye contrast sensitivity or high-contrast Snellen visual acuity using Wilcoxon testing. There was a report in New England Journal of Medicine that commented on the relation between concussive tympanic membrane rupture and blast-TBI in military casualties in Iraq, so we evaluated that. We did not find a significant correlation between the two; however our sample with ruptured eardrums was relatively small. We found a very, very low correlation between any protective eyeware use and ocular injuries. We have patients who have the raccoon eyes from impact of the eyewear frames, and a peppered face indicating exposure to a cloud of fragmentation. Despite the obvious use of eyewear, some had internal eye injuries that were significant.

Getting away from the zone discussion, the most common injury of all of their injuries is ciliary angle recession. I think 30%, or one out of three, of this group from the blast has angle recession. It is a marker for significant damage to the iris and a predictor for traumatic glaucoma in the future. What is a follow-up going to be for this group of people? What do they have, and what are their needs for ongoing care? We also some subjects who have required oral medication and topical medicines to control glaucoma.

Quality of Life

The other thing I want to talk about is quality of life. The two areas I want to cover here are: the Visual Functioning Questionnaire (VFQ) and the Neuro-Ophthalmology Supplement to the VFQ. The National Eye Institute developed the VFQ 25. A score of 100 is perfect; zero is terrible quality of life. Various studies have validated these questionnaires. There are numerous vision domains, including distance acuities, near vision, peripheral vision, and others. There are a few areas that we see that are not captured by the VFQ 25. Other elements captured by the Neuro-Ophthalmology 10 include double vision, eye movement, photosensitivity, and ptosis. Here are the results of our statistical analysis. I won't bore you with it unless you are interested.

Our distribution of our VFQ 25 was pretty widespread from 30% to 100%. They did not all rate themselves perfect or terrible. There was a little bit of difference on the neuro-ophthalmology supplement. What we found when we looked at the composite scores for TBI was that they certainly scored worse than a reference level, which is normal, but they also scored worse than patients with known multiple sclerosis—and neuropathy. It was surprising that they scored themselves so lowly. A trained nurse administered all of these tests. In all cases, we used the same nurse, same format, same room, and the same questions. With the Neurological Outcome Scale (NOS) we found our TBI scores were worse than in the disease-free reference group as well as worse than the multiple sclerosis cohort, but not at a significant level. The subscales were highly correlated.

Interestingly, monocular patients tended to score better on the NOS than binocular patients. They lost half their visual field, but perhaps they do not have the binocular disorders that Dr. Goodrich was talking about. , We could not find a correlation between ocular injury and visual quality of life.

Now contrast sensitivity and quality of life. We do spatial contrast sensitivity on everyone and found it reduced. This is kind of the scatter plot that we have with our particular machine. On the CSV-1000—spatial frequency the highest score is a log 2. That would be to get all 8 correct. All eight choices at that frequency. You can see a trend here to correlate with quality of life, which is significant.

Automated perimetry: we are very interested in the mean deviation and the standard pattern deviation that is well validated in glaucoma studies. I took those with glaucoma out of the analysis. Excluding these we still had a significant mean deviation, but even more with the standard deviation and the visual field scores. Again, we did not find a correlation so much with TBI severity level. This is what the bar graphs look like.

Despite that, we have seen a number of vision-threatening complications while evaluating. We had traumatic glaucoma develop while evaluating patients.

Discussion

>> AUDIENCE MEMBER:

I was wondering if you did any kind of general quality of life testing. I assume you did not.

>> *GLENN COCKERHAM:*

Like an S.F. 36? We have started that, but I do not have it for the original group. I was hoping somebody else from the VA would have done it but, in many instances, it was not done.

>> *AUDIENCE MEMBER:*

How about any soft structural measures like OCT [Optical coherence tomography]?

>> *GLENN COCKERHAM:*

Yes, I just have not analyzed it. We are in the middle of the study. We have analyzed ocular injuries because it was of immediate interest to us. We stopped and looked at the quality of life, but we are still performing the battery of tests. We have not yet looked at specular microscopy and accommodation, for instance.

We continue to see new patients. We had a tremendous surge from Afghanistan that took us through Christmas 2011. We are kind of in a lull now. While the military takes care of their weaponry we stop and analyze.

>> *AUDIENCE MEMBER:*

Glenn, can you amplify the visual field findings and expand on what kind of visual field loss you had beyond the hemianopia?

>> *GLENN COCKERHAM:*

The main thing we found was reduced mean deviation, often without scotomas. We found scattered depression throughout the field. I have no idea if it was due to medication, or headache or something ocular, although there were no observable ocular injuries on thorough examinations. All I know is that fields were depressed.

>> *AUDIENCE MEMBER:*

It is there.

>> *GLENN COCKERHAM:*

And I have seen a number of these in follow-up. The field depression appears to improve with time. It may even go away while the patient is in the hospital. That is cured pretty rapidly in most cases. We have not yet formally analyzed longitudinal changes.

>> AUDIENCE MEMBER:

To what extent, the first time you administer perimetry to somebody, will you get greater noise and greater variability in deviations and standard deviations? Do you have the opportunity to do multiple tests and retest?

>> GLENN COCKERHAM:

In many cases, we do and our result stands up. Sometimes it is better, but in most cases we continue to see field depressions. We will have to analyze the longitudinal data at the end of the study to know for sure.

>> AUDIENCE MEMBER:

Earlier we heard that there is really no validated kind or set of ophthalmic examinations to assess the convergence, the pursuit of disorders part of traumatic brain injury. I was wondering what you are using in your examination?

>> GLENN COCKERHAM:

Our neuro-ophthalmologist performs a complete examination, including fixation, convergence, saccades, and pursuits. Perhaps Kim can talk about that tomorrow. We perform accommodation testing with a Behrens rule during testing of visual function, as well as with dynamic stimulation of accommodation with a wavefront analyzer. We are just beginning to measure accommodation with an infrared analyzer, the WAM-5500. In addition, we perform video nystagmography for horizontal saccades, pursuits, and optokinetic nystagmus.

>> WILLIAM GOOD:

Thank you very much. It is now my pleasure to introduce Dr. Matthew Harper. He is a research scientist in neurobiology at the Iowa Center for the Treatment and Prevention of Vision Loss. He is going to be lecturing us this morning on the topic, new information learned from external blast TBI.

New Information Learned from Experimental Blast TBI

MATTHEW HARPER, PhD

Thank you for the opportunity to share our research results. The goals of our research were to identify the acute and chronic vision problems associated with experimental blast-mediated TBI. We have used functional and structural analysis of the visual system. We have also looked at chronic neuromotor and histological consequences of TBI. Our ultimate goal is to identify molecular targets we can use to develop treatments for TBI. In our experiments we have developed a blast chamber for mice. Half of the chamber is pressured and is covered with a small Mylar membrane that bursts at 20 PSI. A blast wave passes over an anesthetized mouse.

Retinal Deficits

We wanted to start by looking at visual function in these mice. Briefly, if we shine red light in the mouse's eye, it primarily activates photoreceptors. If we shine a blue light into the mouse's eye, it targets a sub population of photoreceptive retinal ganglion cells. We examined at the maximal pupil constriction in mice after TBI with a video camera. In our TBI animals we can see that the pupils do not constrict as well as in our control animals. We also wanted to look at the baseline pupil diameter. We did not see significant resting pupil deficits. In a normal animal that is in the dark and has about a 2-millimeter diameter pupil, it constricts down 1 millimeter with a red light and constricts to 0.5 mm with the blue light.

We wanted to investigate the potential visual deficits further, so we used the pattern electroretinogram (ERG). A white and black line is the standard rodent pattern ERG stimulus to examine retinal ganglion cell function. We see about a 5 microvolt response to our stimulus and can see that there are significant pattern ERG deficits. One year after TBI the mice retained this significant pattern ERG deficit. So we used optical coherence tomography to look at the retinal nerve fiber layer of the ganglion cells to see if they were affected.

We looked at control mice versus animals with TBI in a couple of different ways. We either included the blood vessels or excluded them. Both methods reached significance three months after injury. There is a chronic degeneration of retinal ganglion axons after one exposure of mild TBI.

26

Behavioral Deficits

We examined what else is affected in addition to the visual system in mice with TBI. To answer this, we looked at coordinated motor activity using the Rota Rod analysis. The Rota Rod is a spinning rod that mice like to run on. If they do not have coordinated motor function, they will fall off and stop a timer.

Control mice ran for 20 minutes on this Rota Rod. Seven days after injury we can see that there is a significant decrease in coordinated motor function, which actually resolves by three weeks after injury. There is some form of spontaneous resolution of this coordinated motor activity deficit. It appears to be consistent if we look at any farther than this 21-day period. These animals then perform normally at both running and walking speeds.

We used a laser grid system to look at the horizontal activity in the light and in the dark. What we found was that there was not a significant difference between animals that had traumatic brain injury one and three weeks after injury. We found no significant difference overall activity. They did, however, had very acute deficits and vestibular motor activity.

After this study we decided to look at the cognitive consequences of one exposure to one blast exposure. To do this, we used the Barnes maze, which has 36 holes within a circle. Mice are very agoraphobic; if we lift the containment box, the mice learn to run into this escape box because they like being in a secluded area. When we looked at animals that had TBI versus the control animals we found that the animals with TBI took much longer to learn the task as compared to the animals without TBI. At the end of the experiment, we looked at the cumulative latency to the target. The animals with TBI took much longer to find that target, indicating that they had difficulty in learning that task. In addition, they made many more errors to the target. An error to the target is going to the wrong hole. At the end of the experiment we saw that there was a significant increase in errors in our TBI animals, which led us to ask if blast exposure affect pre-learned tasks.

We trained a group of animals over a period of ten days and then induced TBI in these animals. We started testing these animals 1 hour, 5 hours, 12 hours and 4 days after the injury and showed in fact that these animals retained the previously learned memory regardless of the delay period. TBI did not affect memories learned before the injury. However, it is very difficult to teach the animals new memories after TBI.

This led us to perform some histological analysis of this tissue in the thalamus. We are using a green dye that permeates neural membranes that are compromised. We looked at the control animals and there is not any staining in those animals. By seven days after injury there is a tremendous increase in the number of thalamic neurons dying and it starts going away by 21 days. Importantly, 21 days is when we see the spontaneous recovery in the coordinated neuromotor function. This may be due to some form of neuronal plasticity.

Histological Analysis

We looked a little bit further to try to identify some targets that we could use for treatment of TBI.

Using a focal polymeras chain reaction (PCR) array, we found an increase in inflammatory-related genes compared to our control animals. We have a lot of chemokines and their receptors upregulated. We did some histology and we found that lial fibrillary acidic protein (GFAP) was up-regulated starting at 5 hours after injury, which progressed to one week after injury.

We also showed that there is a tremendous amount of oxidative stress in the animals. We looked at the thalamus and the retina. We used 3-nitrotyrosine (3NT), 4-hydroxy-2-nonenal (4HNE) and inducible NO synthase (iNOS) proteins to assess the oxidative stress response. There is an increase in each of these after TBI. What is interesting in these results is that neuron tracks appear to be lighting up with iNOS. These are all from the lateral geniculate nucleus (LGN), in the thalamus. One is a representative example from the 3NT expression, a stepwise increase up to about one week that the oxidative stress molecules are increasing.

Probably the most important finding is that we are finding neurofibrillary tangles using two different antibodies. In the control animals we do not see any neurofibrillary tangles, compared to extensive antibody staining in animals with TBI. About 80% of the mice expressed these neurofibrillary tangles are all over the brain.

Our conclusions so far are that with the mild TBI we have a decrease in coordinated neuromotor function, although the overall activity of the animals remains consistent. Memory acquisition, but not retention is affected. Focal damage is present throughout the brain where we have up-regulation of proteins associated with oxidative stress and inflammation. These oxidative stress and inflammatory molecules can interact with each other and cause oxidative stress, inflammatory cycles, which increase neuron degeneration.

Perhaps the most important finding we found so far is that this single blast injury results in chronic neurodegeneration one year after the injury. These mice essentially look like a normal mouse behaviorally, but when we take the brains out of them and look at them, they look like mice with Alzheimer's disease.

I acknowledge the VA and the Iowa Center for Advanced Neurotoxicology for their funding.

Discussion

>> WILLIAM GOOD:

Questions? I will start with one.

This is a great model for understanding the acute and chronic vision problems of our TBI patients. Do any of your experiments look at manners of protecting the brain after the blast, any thoughts that come to mind on that score?

>> MATTHEW HARPER:

We intend to use some type of anti-inflammatory or antioxidant therapy in addition to some protectant therapies that try to prevent neurofibrillary tangles in the neurons.

>> WILLIAM GOOD:

Another question is: if you can manipulate the genome of the mouse so easily, it seems that this model would also lend itself to experimenting with different types of mice. Yes?

>> MATTHEW HARPER:

Well, we so far looked at one type of mouse. We looked at a mouse that is pre-disposed to develop Parkinson's disease. If you give that mouse brain injury, it will develop Parkinson's disease 50% faster than animals not exposed to the disease. This is an observation that suggests soldiers' chronic health may be severely impacted by the traumatic brain injury.

>> *AUDIENCE MEMBER:*

How did you arrive at the force that you used that equated to the harm of the blast?

>> *MATTHEW HARPER:*

It was a little bit of luck actually. Previous studies have shown the LD 50 for rats was about 30 PSI. We got a hold of some Mylar membrane and ran it through our blast tank. It turned out that the membrane burst at 20 PSI. We had some engineers do calculations. They found that at about 6 meters from IED blasts there is a 20 PSI blast. Soldiers are often times exposed to blasts in this range.

>> *CHRISTOPHER TYLER:*

You characterize the damage as diffuse but focal. I'd like to resolve that conflict. When you say "cortex", there are different ways to analyze cortical damage, white matter damage, subcortical damage, brainstem damage, and so forth.

>> *MATTHEW HARPER:*

We have looked at the cortex so far. We are going to go back and do a further stereotaxic analysis so that we can actually quantify where these neurofibrillary tangles are. When I said diffuse, I meant that this damage is present throughout all fibers of the brain. The focal degeneration is really in the retina. We see these areas where ganglion cells are dead and dying. There are areas with ganglion cells and areas that are absent of ganglion cells.

>> *CHRISTOPHER TYLER:*

I'm assuming that you did not do this with the hippocampus.

>> *MATTHEW HARPER:*

We have not looked in the hippocampus at all yet.

>> *AUDIENCE MEMBER:*

I agree with Dr. Good. This is fascinating and disturbing evidence. During these experiments, the mice are anesthetized. Obviously, in the battlefield, soldiers are not under anesthesia. Do you have any data on the how the affect of anesthesia would influence the brain and the retina in regard to the blast?

>> *MATTHEW HARPER:*

The question was: What is the affect of anesthesia? Some can be neurotoxic, so we switched to gas inhalation. It is 4% halothane and runs into the mouse continually. The mouse wakes up from the anesthesia in one minute. So, I think that the affect of anesthesia on the mouse is minimal for the brain injury. But I know that there is no way to judge, we really cannot test this.

>> *PIA HOENIG:*

Once you saw the neuronal damage for the animals, did you try other parts of the brain? At the Muir Lab we just started looking at the thalamus, so what was the most likely target for you?

>> *MATTHEW HARPER:*

There is neuronal damage that had occurred in the cortex, which mirrors that of the thalamus.

>> *AUDIENCE MEMBER:*

You showed that the blast affected learning after the blast. Was there a learning curve? Did the mice find the exclusion box eventually? And then, later on, when challenged to repeat the search did they then forget more easily? Was there plasticity in learning? And then, was there degradation of that plasticity or recontinuation of that plasticity?

>> *MATTHEW HARPER:*

Those mice had difficulty acquiring memory; but once the mice acquire the memory, there is plasticity. We let those tested sit alone for three months, and then we came back and analyzed them. They performed fine on that maze. For those mice that had that deficit, that was three months after injury. It is quite a while.

>> *WILLIAM GOOD:*

Any other questions?

>> *AUDIENCE MEMBER:*

My question is about the strength of the blast. Do you have any idea as to whether it would coincide with what we would consider mild, moderate or severe TBI?

>> MATTHEW HARPER:

We think that it is mild TBI because when we look pathologically, especially in air filled organs, we see a little bit of damage in those. We obviously see damage in the lungs, but not extensively. We are not seeing hemorrhages. We are seeing vessels that are leaky with Immuno-gold (IG) staining. I think this is a good model for mild TBI because we do not see the larger gross deficits.

>> WILLIAM GOOD:

Our next speaker is Gabrielle Saunders. She is an investigator and deputy Director for Education with National Center for Rehabilitative Auditory Research and a Professor of Otolaryngology, both of which are important. She is doing a lot of research on auditory rehabilitation for mild TBI.

Impacts on Hearing

GABRIELLE SAUNDERS, PhD

Throughout the morning we've heard about the multi sensory problems that can occur in soldiers who are subjected to blast injuries and concussions. My job now is to tell you about damage that occurs to the auditory system.

What I find fascinating is that so much of what has been said about TBI and the visual system has analogies in the auditory system – in terms of what we know and do not know, the research gaps, and the types of symptoms reported. On that note, just before I get onto the main part of my presentation, I want to remind you that these two sensory systems (visual and auditory) are not independent – especially when it comes to higher level processing in the brain, and as a result, there are many individuals with blast-related mild TBI who report both auditory and visual difficulties in the absence of damage to the peripheral organs (eyes and ears). If we are to really understand what is happening following blast exposure, I think it is a mistake for us to look at each system independently. In fact, there is a huge gap in knowledge about dual sensory impairment and the impact it has on daily function and quality of life. It is assumed that the impact of dual sensory impairment, relative to single sensory impairment, is multiplicative rather than additive, but really this has not been well investigated. Of course, the presence of dual sensory impairment also has huge implications for rehabilitation. Once again, however, this is rarely considered when selecting rehabilitative interventions. There is lot more I could add on this topic, but right now I should get on and address what I was asked to talk about - ways in which blast exposure impacts the auditory system.

I am sure you are all familiar with the textbook image of the ear. One 'interesting' thing about these images is that they fail to convey the fact that hearing is more than merely the detection of sound – it requires central auditory processing to make meaning out of the incoming auditory signal. The ear alone cannot localize sounds, extract meaning from speech, focus attention or recall information. In other words, once the auditory signal has passed through the outer, middle and inner ears, it passes up the auditory nerve via the brainstem to the brain, where auditory processing takes place. If anywhere between the pinna and cortex is not fully intact, auditory communication will be disrupted.

So, what does blast do to the auditory system? Starting at the outer ear, the pinna can be burned or damaged by flying debris. The tympanic membrane can rupture, the ossicular chain can become disarticulated and the ossicles can be fractured by the high-pressure wave generated by the blast, resulting in conductive hearing loss. The inner and outer hair cells on the basilar membrane can be torn from their support cells, which leads to hair cell death; and also possible, but less common, is damage to the semicircular canals, causing vestibular (balance) problems and dizziness.

Beyond this, recent work strongly suggests that blasts cause central auditory processing difficulties. This is presumably due to the damage that occurs when the brain moves within the skull causing diffuse axonal injuries (shearing and stretching of neurons), contusions and subdural hemorrhaging. Taber et al. (2006) have shown these types of injuries are seen in areas of the brain necessary for auditory processing (temporal, parietal, and frontal/prefrontal cortices, corpus callosum and thalamus).

What is the result of these types of injuries? In order to answer this question I will present data from two recent studies conducted at the National Center for Rehabilitative Auditory Research (NCRAR) in Portland, Oregon, where I work, along with other collaborators.

The first study, the findings of which are now published (Gallun et al., 2012), was a collaboration between the NCRAR and Walter Reed Army Medical Center in Washington, D.C. The study compared the performance of 36 blast-exposed patients and 29 non-blast-exposed subjects on a battery of central auditory processing measures and electrophysiological tests with a view to determining whether central auditory processing abilities are impacted by blast exposure. All individuals in the study had clinically-normal hearing thresholds, however the blast-exposed group did have significantly poorer pure tone sensitivity than the non-exposed group, and 78% of them complained of more difficulties hearing in noisy environments following exposure to blast. The groups did not differ significantly on speech understanding in quiet but more blast-exposed individuals performed outside normal limits on a test of speech in noise (the QuickSIN test; Killion et al., 2004) than non-blast-exposed individuals. This suggests that, although speech understanding *per* se was unaffected by blast, listening in a more complex auditory environment is impacted.

The behavioral test battery completed by all participants consisted of a measure of temporal pattern perception (the Frequency Patterns test, Musiek et al., 1989), a test of auditory temporal resolution (Gaps-In-Noise test, Musiek et al, 2005), a measure of binaural processing (Masking Level Difference), two tests of dichotic listening (Dichotic Digits Test; Musiek, 1983), and the Staggered Spondaic Words Test (Katz, 1998). In addition, auditory brainstem responses (ABRs) to assess integrity of the cochlear and central auditory pathways between the auditory nerve and the superior olivary complex were assessed, as were long-latency electrical responses (N100, P160/P200, N200 and P300) using an oddball paradigm. These measures assess attention, cognitive and auditory discrimination, memory and semantic expectancy. For more details about the methods used see Gallun et al. (2012).

The data were examined by comparing the percentage of subjects in each group who performed more than two standard deviations outside the mean of the control group. More blast-exposed individuals than non-blast-exposed individuals performed 'abnormally' on the Gaps-in-Noise test, Masking Level Difference, and Staggered Spondaic Word test, but not on the Frequency Patterns test or Dichotic Digits test. The total number of tests on which individuals performed abnormally was then examined and plotted. As you can see, the proportion of blast-exposed individuals performing abnormally on one or more tests is much greater than the number of non-blast-exposed individuals. Specifically, three quarters of the blast-exposed individuals performed abnormally on at least one test, as compared with only one quarter of the non-blast-exposed individuals. It is important to note that none of the participants performed poorly on all five test measures as it negates the concern that subjects had a global cognitive deficit or that they were not making an effort during testing. Looking at the electrophysiological data it was seen that both groups had normal and similar ABR latencies and amplitudes, while the longer latency responses showed some group differences. Specifically, the blast-exposed group had significant longer latencies and lower amplitudes for the P300 response in the right ear.

This study then, shows that blast exposure does seem to affect performance on a number of central auditory processing measures, and specifically those measures that require higher level cortical processing.

What are the resultant complaints these individuals report? Well specifically, they report problems hearing in background noise, following rapid speech, following instructions, and following long conversations.

They also report tinnitus and hyperacusis (hypersensitivity to loud sounds). It is interesting that these are not acuity-based complaints, but are more associated with auditory processing problems.

How many individuals are affected? Well, here are some statistics: About 300,000 Operation Enduring Freedom (OEF)/Operation Iraqi Freedom (OIF) Veterans have some form of traumatic brain injury (TBI), and about 75% of those TBIs are due to exposure to blast (Owens at al., 2008). Sixty six percent of Veterans with deployment-related TBI and blast exposure complain of auditory difficulties and of these 35-54% have sensorineural hearing loss, 7% have conductive hearing loss in the form of a ruptured tympanic membrane and 20% have 'normal or almost normal' hearing sensitivity (see Saunders & Echt, 2012, for a summary).

The big question is what to do for these individuals. Back in 2009 we sent out a survey to VA audiologists asking them how many OEF/OIF Veterans they saw monthly who had normal hearing sensitivity and yet reported hearing difficulties. Of the 89 individuals who replied to the survey, 35 reported seeing four or more such individuals each month, and 47% reported seeing 1-3 per month. In other words, over 90% of audiologists were seeing at least one individual a month who required some form of auditory rehabilitation that, in light of normal hearing sensitivity, could not be addressed through provision of hearing aids. Given that the problems encountered by these Veterans impact daily function and quality of life, the need for rehabilitation cannot be ignored. In my presentation this afternoon, I will address this. Thank you.

Panel Discussion

William Good MD (Moderator), Gregory Goodrich, Glenn Cockerham, Matthew Harper, Gabrielle Saunders, Stephen Heinen

>> AUDIENCE MEMBER:

What kind of therapeutic interventions do you use? Or what were the main ones?

>> GREGORY GOODRICH:

I'd like to reiterate your point that we should not separate vision and hearing too much. Henry Lew and a group of us looked at a retrospective record review of patients in our polytrauma rehabilitation center (Lew et al., 2009, 2010). Over a third of them had both hearing and vision loss that were impacted the most. When we looked at Functional Independence Measure (FIM) scores, it was the dual loss that was reported as reduced both at admission and at discharge. It was not those with sensory hearing or vision loss only. In the population that we looked at, only 15% of the population had no hearing loss, vision loss, or a dual sensory loss. Loss of vision, hearing, or both affected 85% of the population.

>> GABRIELLE SAUNDERS:

It does not surprise me that having both hearing loss and vision loss is cumulative. We rely on the second sense when we cannot use the first. Rehabilitative interventions use this approach too.

>> AUDIENCE MEMBER:

It seems that the third end organ is smell. Have you evaluated that?

>> GABRIELLE SAUNDERS:

No.

>> AUDIENCE MEMBER:

As a clinician, there is no question that loss of both audition and vision is a synergistic phenomenon. Vestibular takes you to the next step into oblivion.

What is fascinating to me is that we are talking about young kids with TBI. We have talked about fibrillary tangles. A lot of our elderly people who are

37

having problems with their hearing and are probably not assessed appropriately either. Simply offering hearing aids is probably missing a majority of their deficits as well. This work has a lot of extensions into the non-traumatic population.

>> GABRIELLE SAUNDERS:

I entirely agree with you. The typical clinical assessment measures focus on threshold listening, which is clearly not typical of real world listening, or single word recognition – also not real-world listening. Furthermore, I think it very unfortunate that we do not (yet) have any assessment tools that require the integration of both hearing and vision. While such tools would be difficult to develop, I believe they are critical for evaluating the impacts of the kinds of problems we have been talking about today.

>> AUDIENCE MEMBER:

We see localization, identification of an item of interest, whether it is visual or auditory, and the ability to discern speech with impaired vision perception. I have had patients who became disabled because their hearing dropped, but their fundamental problem was that they were legally blind.

>> GABRIELLE SAUNDERS:

Yes. There are no clinically standardized measures out there. I would love to advocate for some.

>> YURY PETROV:

It looks like the strongest affect was in the task that requires attention. How specific is it to hearing?

>> GABRIELLE SAUNDERS:

That is a good question. Since there are data out there from blast-exposed Veterans with normal vision who report problems with sustained reading, it seems highly possible that there are common underlying causes that might be associated with attentional deficits. However, in light of the many other tests on which the blast-exposed individuals performed abnormally, I doubt attention is the sole underlying factor.

>> YURY PETROV:

In the digit test, do they allocate attention?

>> GABRIELLE SAUNDERS:

If you mean, are they directed to focus on one ear or the other, the answer is no. They are just instructed to repeat back as much as they can.

>> WILLIAM GOOD:

We'll have a chance for more questions, but why not come and join the panel.

>> WILLIAM GOOD:

Joining this panel now is Steve Heinen.

I will start the questioning off with a question I have: My understanding is we now have men and women in combat. My question is: Has anybody looked at gender differences in the responses of people to a blast injury or mTBI?

>> GABRIELLE SAUNDERS:

Not to my knowledge. In the studies I reported above there were only seven women so the data don't lend themselves to that level of analysis.

>> AUDIENCE MEMBER:

In the earlier discussions there was some references to the types of brain injuries you get that tend to be to more central brain structures and sort of multifocal. My question for Dr. Saunders and all of your data concerns looking at correlations between different deficits and how they relate anatomically. You mentioned several tests representing different portions of the central auditory system and having anatomic correlates. One question is: Are there people who are abnormal on one test that are likely to be abnormal on another test? Clearly no one test will put it all together, but have we started to put together cross-correlations in terms of deficits to start saying, if you have that one you are likely to have this one?

>> GABRIELLE SAUNDERS:

The authors of the study I described did look at correlations between performance on each measure and found generally low correlations. Now, because they specifically selected the test measures to examine different levels of auditory system function this is perhaps not surprising.

In terms of which tests to use in the clinic – from a practical standpoint, it seems to me that conducting multiple tests to determine a specific site of the lesion is somewhat premature until we have rehabilitation that can target specific areas of brain deficit. Right now we have limited rehabilitative options available to address central processing deficits. It therefore seems to me (although I know some would disagree) that if the rehabilitation to be provided is the same regardless of the site of lesion, why not stop assessment testing once *any* deficit is noted. I am not sure it makes sense from either the clinician's or patient's perspective to spend a lot of clinic time conducting tests from which the results will lead to the same rehabilitative intervention.

>> WILLIAM GOOD:

I have an ophthalmic question for Glenn. You talked about open and closed globe injuries. Do you see sympathetic ophthalmia in the field or, are there things you do to try to prevent it, or is the incidence so low you do not worry about it?

>> GLENN COCKERHAM:

There have been two cases that may potentially have been sympathetic ophthalmia since the war started in 2001, but this is not clear. Dr. Dan Elizonda just came back from six months duty as the military ophthalmologist in Kandahar, Afghanistan. The ophthalmologists in Afghanistan routinely did enucleations in the theater, before sending patients downrange to the U.S. hospitals.

The prevention is to take the eye out before the inflammation occurs, but I have not seen war-related cases myself. Dr. Mazzoli perhaps could address it better. I do not know of any documented cases at this point.

>> ROBERT MAZZOLI:

I am only aware of one semi-documented case of sympathetic ophthalmia. I would be interested to know if there is a second case out there. Throughout the war, we have had only one case that has been documented.

>> GLENN COCKERHAM:

Through the duration of the war, we had one published case of sympathetic ophthalmia that was eviscerated in the theater in Iraq. It was diagnosed as sympathetic at an academic institution in the Midwest. This case was treated fairly quickly with a very short course of oral steroids and topical steroids and reportedly cured.

Whether that prompted the diagnosis, as I understand it, was based on a few cells - some early spillover into the posterior chamber. Whether that truly was sympathetic or not, we do not really know. But we are treating it as sympathetic, nevertheless. To my knowledge this is the only case. It was published in the Ophthalmic Plastic and Reconstructive Surgery Journal, in part because a leading reconstructive specialist wrote it.

>> *DONALD GAGLIANO:*

Regardless of who made the diagnosis, we are not questioning it at all. As far as we know, of the open globes we have had only one well-documented case of endophthalmitis. This goes to what Glenn is saying about primary nucleation. There may be some other confounding factors, but the answer is surprising given the number of injuries we are seeing of open globes. We have one documented case of sympathetic.

>> *WILLIAM GOOD:*

Very interesting. For those of you who do not know what sympathetic ophthalmia is, it is an autoimmune reaction that occurs in the normal fellow eye after the other eye has been injured.

>> *DONALD GAGLIANO:*

What I find amazing is that with all the retained intraocular foreign bodies we have only recently had the recent case of endophthalmitis from all the open globes. I wonder if this has to do with the nature of the injury, the hot blast objects injuring the globe? If it has to do with the nature of the injury and hot objects perforating the globe as opposed to elsewhere, vegetable foreign matter.

Typically in conventional modern warfare, it is hot pieces of metal shrapnel that self-sterilize and you might think these are a metal, but sterile intraocular foreign body. What we find with the nature of an IED is that the bomb is typically encased in metal so there are metal foreign bodies. Sometimes the weapon is buried in the rock, in the gravel, so very often you will find gravel foreign bodies. If it is a case of a suicide bomber, and this is grotesque to think about it, sometimes you actually find bone or flesh, or mud is the intraocular foreign body. Sometimes those foreign bodies are bits and pieces of bystanders or the suicide bomber himself. Now, you have human, mammal carcass, rock, sand, all matter of vegetation, clothing, intraocular foreign bodies.

So despite the broad, broad spectrum and nature of intraocular foreign body we still have had only one endophthalmitis, and that went rapidly within 48 hours of injury.

>> WILLIAM GOOD:

It is my understanding that the speed of the projectile as it hits the tissue is important. The projectile has to be traveling at Mach speeds as it hits the skin or cornea. At that speed then it is sterilized just from the impact?

>> DONALD GAGLIANO:

The medics are now trained to act rapidly if there is an open globe

>> ROBERT MAZZOLI:

Medics are now trained to act rapidly. One of the things that we also do within a matter of minutes to hours, if there is an open globe or a suspicion of an open globe, is administer an antibiotic at the site of injury. The idea is that they are getting systematic antibiotics that are broad spectrum early on and that antibiotic is continued through the course of evacuation.

>> RANDY KARDON:

I have struggled with the nature of the injury as it relates to clearly differentiating between ocular damage and traumatic brain injury or actual brain damage, for example, with animal responses. Is it damage to that system in the eye, or is it damage to processing pathways further down?

Glenn, do your results show if is it damage to the eye or is it visual information processing pathways, where we are getting those overall symptoms that we see?

>> GREGORY GOODRICH:

We requested early inquiry if we are looking with OCT and retinal imaging and we're doing it serially. If we can nail it down in the visual pathways, I think the retinal ganglion cells are affected. It could be cognitive or it could be drug-related. It is hard to sort this out. Hopefully the Cirrus HD-OCT, which we are using now, will give us clues. We tried Stratus, and I was not very satisfied with that. We also tried HRT and then dropped those two tests. We are now sticking with the Zeiss because it is all we have.

I am not able to answer the question: where is accommodation affected? Is it brainstem or iris or ciliary body or what? I do not know. Maybe this is a direction for analysis. We have not evaluated this area yet and maybe it will give us some clues.

>> AUDIENCE MEMBER:

The story about the pupillary eye reflex is interesting. As I understand it, if you control the way that you give the stimulus in terms of this state of adaptation you can tease out whether it is afferent or efferent. In other words, it depends on whether you are using a red or a blue light, and if you're recording both pupils simultaneously.

Medtronics has a device that measures pupillometry from concussive impacts and also intracranial monitoring devices for the monitoring the inter-cranial depression. It is called a neural pupillary index and it measures several different parameters for how the pupil responds. They found that it was affected even before the intracranial pressure started to go up. The brain edema can affect the dynamics of the pupillary light reflexes in a predictive way. One of the studies is in the emergency treatment room where we have four trauma bays. We are collecting light reflex on every patient to see what happens to those patients. Do they have deficits? Can they go home? Is there something predictive about the efferent defect?

>> RANDY KARDON:

We are doing a lot of studies on the pupil as a screening tool for diabetic retinopathy. In our studies on light sensitivity we are looking at using diphtheria toxin to knock out inner ipRGCs. You have to actually wipe out all of them before you lose the pupillary response. When you see that dramatic change in a fairly significant ganglion loss, it probably is not solely due to that. You almost have to knock out the entire population before you lose any kind of pupillary response. There is undoubtedly a central component to it.

>> DONALD GAGLIANO:

Is there concussive injury in your model? Or is it purely blast wave? In a lot of the actual patients there is also a component of the concussive injury from the blast wave that causes some impact, this is another element of the injury. I think that your model is really an injury that requires that you completely eliminate that coup/contrecoup. Is there a coup/contrecoup in your model?

>> RANDY KARDON:

There probably is. I say this because the pressure wave, when it travels through the tissue, it is going to cause movement of the brain within the skull but not nearly to the extent of the concussive injury. Matt alluded to this. We can think of it as a blast and a concussive injury, if you take some of the padding from around where the mouse is.

>> ROBERT MAZZOLI:

I have a question for Glenn and Dr. Saunders. I saw the three-line curve and I saw your blast physics, Friedlander curve. This made me wonder about Glenn's comment that wearing eye protection was not related to the presence or absence of true ocular damage. What is the blast physics behind the eye, between the back of the eye protection and the front of the eye? We know that open air blasts may follow a Friedlander curve, but behind that what goes in rebounds and changes in that blast wave. Has anybody started to look at the actual model of the blast wave behind eye protection and the eye?

>> GLENN COCKERHAM:

I believe we are finding things at the Palo Alto VA related to the problem. I did not mention the specular microscopy data.

Certainly this is an issue for the individuals who wore the Wiley X or the Oakley M protective frames because they are pretty flexible. These devices can be distorted to flex with a strong concussive shock wave, leading to physical impact of the back of the lens or the compressed air behind the lens on the eye. We certainly can see very reduced endothelial cell counts in individuals who were wearing eye protection, because I can see the outline of the frames on their periocular skin.

In many instances the glasses are actually impacted into the skin. It leaves a scar where the edge of the frame went into the skin. We know they were wearing protection, and yet, they have these internal eye injuries. It certainly looks obvious to me that something is hitting and indenting that cornea and also causing angle recession.

>> WILLIAM GOOD:

You've got to believe that we'll be seeing a lot of cataracts at some point.

>> AUDIENCE MEMBER:

Who interacted with these patients? Even though you talk about low-level damage, it sounds like the patients themselves they do not complain.

Rather, it seems their complaints are more about high-level deficits. You mentioned they had trouble reading. They tire quickly. It sounds like they may have some high level problems that are not necessarily related to vision or auditory processing sustaining their attention.

What are your own feelings about this possibility? If you had to test them, would you rather pay more attention to the high-level abilities or more low-level functions? 20/20 vision probably does not matter too much to them. Losing the ability to focus and concentrate is probably much more important for them.

>> GREGORY GOODRICH:

If someone comes in and says that they are having trouble in school or that they are having binocular issues, we will try to resolve those. But, ours is not a concerted effort to look at all the individual problems. For example, I rather wonder about our neuropsychology assessments, which are very heavily paper and pencil-oriented. If you have people with significant binocular dysfunction, doing a paper and pencil task is their overall rating reduced due to the dysfunction, neuropsychological impairment, or both?

In the underlying research designs and in the questions we seek to answer, we try to scientifically parcel out the effects of an injury to a specific part of the brain, the eye, or hearing etc. However this can be difficult. If we look at the troops we are serving a lot is going on for them.

These people are coming back from deployments. One patient was telling one of the employees at the VA that he spent eight months in Iraq in the Green Zone. You would think that life is relatively good in the Green Zone since it was a large secure area in Baghdad. It probably was, except for the in-between times with the mortar rounds being lobbed into the compound. So even "secure" deployments have hazards.

There are just a lot of psychological effects as well as the physical injuries that we see not to mention those that we are unable to see. We have to go hunting for the latter. And then if we find them, we have to understand them in the context of each individual patient.

You are asking a terribly important question. Practically, it is very difficult to partial out all of the effects, let alone the interactions among them. Hearing and vision, I think, is a great example. Having those two systems affected does reduce functional performance, as opposed to having either one or none.

Functional vision is affected by memory, by attention, by cognition and so on. One of my fears is what we are reporting as vision may be something else in the central processing system. We have to just be aware of that and try to be as cautious in interpreting as we can. Nevertheless, we find strong associations among the sensory losses.

I cannot answer your question, but I can personally underline that you are raising an important point to partial out as we look at what is going on. Rather than saying it is this or that, we probably should say that it may be this, that or those.

>> GABRIELLE SAUNDERS:

From a practical clinical standpoint, to some extent what we want to test for needs to be limited to what we can provide in terms of deficits of central processing. You can do vast test batteries for detailed diagnosis but our rehab intervention is rudimentary. I am not sure it makes sense to spend hours in the clinic understanding where the deficit is if we do the same intervention ultimately.

>> AUDIENCE MEMBER:

As I was listening it became clear that you have patients that we may not have rehabilitation to serve. Maybe it is a Pandora's box? The implications of not having enough rehabilitation are huge. Services and support is huge, absolutely huge and I guess my question for those of you who are the experts in the military system, is: how do you go about defining the parameters factored into the definition of disability when it is used for the people who made these sacrifices on our behalf?

>> AUDIENCE MEMBER

A historical parallel may help here. In the early 1930s, legal blindness was defined as 20/100 or worse: Visual field not greater than 20%. That definition had profound social implications because in itself it opened the door to some misinterpretation. On the original acuity chart you can read the big "E" line and the next line down was 20/100. There was a big gap in that chart that was reflected in that early definition. That definition, however, was established for a defined purpose. When everybody and his brother adopted legal blindness as the definition of blindness it had an impact on the eligibility for services and so on.

So I think parts of what is inherent in this workshop, this symposium, is exploring the definition of disability. As new kinds of injuries emerge and our ability to detect them emerges, how we define things is important.

I think these kinds of injuries were probably prevalent in World War II and in Korea. Maybe not at the same rate of occurrence; I do not know. The difference is that at the time we did not have the ability to detect them. Now we have this ability, so we are looking at science and looking at policy.

I do not think that this group wants to get into policy. Nonetheless, it is something to keep in mind as we consider the evidence that comes out in the next day.

>> DONALD GAGLIANO:

I would amplify and agree with that statement. We certainly are doing a better job at increasing awareness of what to look for and deciding whether or not to treat it. For example, you mentioned these injuries probably did exist in World War II. I would argue that they probably existed in London, as the Germans did a pretty good job of concussively destroying that town, as the allies did a good job of concussively destroying Europe. Yet we do not hear about the number of TBIs. This is not because the German or English physicians were not good observers.

>> AUDIENCE MEMBER:

Was there different name?

>> DONALD GAGLIANO:

We called it shellshock. The difference is that we did not acknowledge it similarly and we were not out there looking for it as aggressively as we are now. Clearly, it existed.

>> WILLIAM GOOD:

Dr. Jampolsky, did you have a comment?

>> ARTHUR JAMPOLSKY:

I think Gabrielle hit on something important here and I wanted to emphasize it. A lot of clinical tasks produce the evidence that has an impact on a practical application on a one-to-one basis. An interesting statement is absolutely true, of course, but what will you do with it about hearing, for example? If you do not understand rapid speech as an older person, is there any magic? Not at all. The magic has been working with noise and speech. It makes the issues smaller, but not more tractable.

>> *GABRIELLE SAUNDERS:*

Incidentally, there is some technology that can improve the signal-to-noise ratio for listening to speech in noise.

>> *ARTHUR JAMPOLSKY:*

We will come to that later. I wanted to point out the differences. With lenses, for example, there are a lot of functional problems and we are not really sure whether these are a part of depression and other things either.

Do you have these same things in hearing? There is a huge difference between vision diagnosis, functional or non-functional and supposedly the vision treatment. Is there is a huge difference between hearing and hearing treatment.

>> *GREGORY GOODRICH:*

Whether we can rehabilitate or not is one very important aspect of this and if we can tell the patient what happened to them. It helps to say: This is not part of your fantasy world. This is not a part of your imagination. This is not a psychic event that happened. We can measure and tell you what components of your sensory system have been damaged. And I think there is a certain peace of mind that comes from patients just knowing what happened to them and regardless of whether we can treat it or not.

>> *AUDIENCE MEMBER:*

I have a question that I am not sure the military and the vet people will like: You have the treatment ability to tell patients what happened in the functional aspects. It is not really clear what you should make of that.

I think a fellow by the name of Ho said in line that there are some different kinds of thresholds for diagnosis. One is what to do when you really do not have evidence that the patient had a blast injury. You just have to say: "you had a blast injury." Is that true?

Can you imagine the hoards of people that are going to line up for this? So if you have an anatomical whatever, the worst case Bob just mentioned, you can have an anterior chamber defect and you are wearing these nice goggles. You do not have any pebbles in the cornea, so you have to stretch to what is implied.

What are you going to do with all these people that are going to line up? The worse thing in the world is to tell someone, "You have a brain injury," when in fact it is a part of depression. That would be the worst, wouldn't it?

>> WILLIAM GOOD:

Well, on that note, maybe it would be a good time to break for lunch. Thanks to all of the speakers and the panel members this morning.

Session 2: Tests, Evaluation and Assessment

mTBI Diagnosis: Objective measures of visual dysfunction
Randy Kardon, MD, PhD

Light Sensitivity in mTBI: Models for assessment and causality
Michael Gorin, MD

Self-assessments and their effectiveness: What vision tests are currently used?
Gregory Goodrich, PhD

Oculomotor function tests, photosensitivity, and convergence testing in mTBI
Suzanne Wickum, OD

Functional imaging for oculomotor deficits in mTBI.
Christopher Tyler, PhD, DSc

Concussion diagnosis
Anne Mucha PT, DPT, MS, NCS

>> WILLIAM GOOD:

Can I have your attention, please? The word has gotten out that we are having a good conference, because the audience seems to have multiplied. Our session this afternoon is entitled 'Tests, Evaluations, and Assessment of mTBI.'

We are very fortunate to have Randy Kardon as our first speaker this afternoon. I have known Randy off and on for a very long time, when I used to travel in neuro-ophthalmology circles. Randy is one of the world-renowned experts on pupils and pupillography. He is a superb neuro-ophthalmologist also. We are pleased to have you with us.

mTBI Diagnosis:
Objective Measures of Visual Dysfunction

RANDY KARDON, MD

Thanks for inviting me. I wanted to thank the organizers, the other speakers and everybody here. It is just the right-size group, and one with good interactions. Thank you.

I also wanted to thank TATRC from the DoD. They are the ones sponsoring a lot of the research work, as well as other branches of the DoD and the NEI. We appreciate all the effort that you go to in executing your programs. You do an excellent job, great thanks.

I want to just hit a few highlights so then we could have something to discuss. I will move through this part fairly rapidly.

Effects of TBI

Most of you have already heard that TBI consists of an initial brain injury, followed by collateral damage associated with a secondary response from the immune system to the acute injury.

Matt Harper, who was our Iowa City VA Center of Excellence Career Development Awardee, presented the animal work on blast-induced traumatic brain injury this morning. One of the interesting aspects of this work that he did not get a chance to mention is that we are finding immune T-cell abnormalities a year out after these experimental blast injuries; that is another aspect of the chronic effects of TBI that we will not have a chance to cover. TBI appears to induce not only an acute inflammatory response of the immune system but also long term effects, which may influence chronic immune surveillance.

I think one of the reasons that so many vision scientists are becoming involved in TBI research is because such a large volume of the brain is devoted to vision and visual processes. Therefore, it should come as no surprise that the patients come in complaining of visualization tasks even though they may have normal vision. There is a great deal of brain real estate subjected to TBI, which can affect visual function.

On a more basic level, one of the most common visual problems seen clinically is the loss of vision in one or both eyes from direct damage to the retina, optic nerve or their connections. Chris Tyler, and some of his

51

colleagues, as well as other scientists on the program will speak more about discoordinated eye movements,

Light Sensitivity

The high prevalence of light sensitivity in patients after TBI is somewhat puzzling. If you survey DoD and the veteran population, 60% of them complain of light sensitivity. We see such patients in our eye clinic at the Iowa City VA and they experience real discomfort in bright light. Still, eye-care specialists who see a patient with light sensitivity often want to walk in the opposite direction, because many are suspected of having a psychogenic problem. Patients wearing really dark sunglasses in an eye clinic often send a mixed message, and many become shunned by health care personnel.

At the end of this talk, I will tell you what we are doing to objectively diagnose light sensitivity, and Dr. Gorin and his colleagues will provide additional results of their studies on the mechanism of light sensitivity.

Diagnostic Strategies

An important strategy to restore sight from blast injury is early diagnosis and treatments that can be implemented as early as possible. How can one assess vision in a patient who has suffered TBI and is unable to respond reliably to standard tests of visual function? In patients who have been cognitively impaired from TBI, one cannot depend on them to be able to perform the same tasks as other patients. If their responses to visual tasks are abnormal, it is often difficult to ascertain whether they have impaired vision or whether their attention is not adequate enough to provide reliable responses.

Yuri Petrov brought up the whole aspect about attentional deficits and how that may adversely affect visual tasks we ask them to perform, which can be a significant confounding factor. One of the solutions to this problem is to use the brain's natural reflexes to interrogate the visual system. Which reflexes are we talking about? There are a number of reflexes that are potentially applicable, but the four that are the most used and most accessible are 1) eye tracking of visual targets, 2) evoked potentials from the brain in response to visual stimuli, 3) the pupil light reflex, and 4) electrical response of the eyelid muscles (electromyogram; EMG) in response to light, known as a photo-blink reflex.

Eye-tracking

Pieter Poolman, who is my co-investigator, is a very experienced engineer in neuro-processing and signal processing. One of the tasks that we have set out to do involves eye tracking to visual targets whose attributes such as contrast and spatial resolution are changed as the target is moved. One eye-tracking system that is uniquely suited to accomplish this is the Smart Eye system. It is a four-camera remote system that derives the gaze angle of the eye based on modeling of each eye's position and head position in three dimensions.

It provides one the ability to use remote monitoring of eye and head position in real time to ascertain a patient's tracking of objects. We are combining eye tracking with the presentation visual "vanishing" optotypes. The optotypes that we use are special, because of the contrast and spatial characteristics of the borders that define them.

Some of you who have expertise in testing infant vision may be familiar with a type of vanishing optotype used in infant acuity charts developed by scientists at Cardiff. It is a binary type of target in that it is either seen or not seen; it is not partially seen. Some of these targets shown on the slide represent a type of vanishing optotype which Dr. Pieter Poolman, my associate, has programmed to present as moving targets on a screen. The rings shown along the top row are decreasing in spatial frequency by decreasing the width of the lines bordering the object and when it drops below your resolution it disappears or vanishes on the background upon which it is projected. Each column shows the same optotype, but decreasing in contrast at a given spatial frequency. When the contrast between the lines outlining the object drops below threshold, the optotype also vanishes against the background. Therefore, spatial frequency and contrast can be independently varied to determine visual threshold.

The idea is you have this type of optotype and move it across the screen at supra-threshold levels of spatial resolution and contrast. As the person is gazing and following its movement, their eye position will correlate with the target position. As soon as is it drops below their threshold, the eye position will no longer correlate with the target position.

Thus, we are using eye tracking as an objective measure to tell whether a person is able to track a moving object that is changing in contrast or resolution during the course of a test. One can imagine how useful this approach would be for infants and for cognitively-impaired people with dementia or after TBI.

Pupil Light Reflex

Another useful brain reflex for assessing vision is the pupil light reflex. Those of you who have some clinical experience, are familiar with using the pupil light reflex in the swinging light test to determine the presence of a relative afferent pupil defect. If one shines the light in the good eye, both pupils will react because they are both connected up in the brainstem. If one alternates the light to the bad eye, then both pupils will not contract as much as when the light was shined in the good eye and in some cases the bad eye sees a decrement of light and will dilate as the light is alternated from the good eye to the bad eye.

The swinging flashlight test is a good objective test to see if there is symmetry of light input from each eye. If there is a bilateral problem, one will not detect it with this test. The test can be made much more precise and accurate if one records the pupillary dynamic movement to the light shone in the damaged eye versus the other eye, which will provide dynamic information to help ascertain if asymmetry of light input is present between the two eyes.

My research efforts have focused on the pupil light reflex for over 20 years. When the role of melanopsin and photosensitive retinal ganglion cells was elucidated in the recent past, that discovery focused the entire field of the pupil light reflex in a new direction. It was previously thought that rods and cones mediated the pupillary light reflex. Genetic models of mice that were produced to study inherited retinal degenerations in the last decade caused complete degeneration of the rod and cone photoreceptors. Histologically, they had no rods and cones but the pupils reacted if a light was shone into the eyes. This reaction was totally unexpected. This led to the search for what retinal neurons other than rods and cones could be responding to light causing the pupil to contract. Eventually, the melanopsin-containing retinal ganglion cells were found to be that neuron.

One interesting aspect about this class of retinal ganglion cells is that they subserve most of the pupillary light reflex to the midbrain. They can be either activated indirectly by input via rods or cones or they can be activated directly by the effect of light on melanopsin pigment contained within the dendrites of these neurons. The spectral sensitivity of melanopsin has shown a maximal response to blue light around 485 nm that has a broad peak. Based on the response properties of the melanopsin-containing retinal ganglion cell, we have pursued strategies to independently assess rod, cone and intrinsic activation of melanopsin retinal ganglion cells with the pupil light reflex. We did this by using a commercially available Ganzfeld electroretinogram unit to produce well-defined stimuli and simultaneously record the pupil response under conditions that favor either rods, cones or melanopsin activation. Thus, the pupillary light reflex was recorded in response to a wide field light at red and blue wavelengths at different light intensities under conditions of dark and light adaptation. The pupillary response was recorded to increasing levels of light intensity to a one-second red light and blue light.

At low levels of light intensity most of the pupil responses are rod-mediated and produce larger pupil contractions to blue light compared to red, when the intensities are matched for photopic equivalence (cones). Therefore, under dark-adapted conditions, with a low intensity blue light, the pupil light reflex is mainly derived from rods. With a bright red light stimulus given under conditions of light adaptation with a blue rod-suppressing background, most of the pupil response is derived from cones. However, the waveform of the pupil light reflex changes dramatically to a one-second light that is in the brighter photopic range at a blue wavelength that activates melanopsin. Under these conditions the pupil contraction is significantly sustained, a characteristic signature of intrinsic activation of melanopsin retinal ganglion cells. Therefore, by adjusting the light wavelength, the conditions of adaptation, and the brightness of the light, one can selectively sample the rod, cone, and melanopsin mediated pupil response. This is extremely useful for sorting out whether a disorder is affecting the rods, cones, or inner retina.

Presently, there are efforts being made by some commercial entities to develop automated portable pupillometers that are designed to make these pupil responses easy to obtain in clinical settings and off-site where telemedical pupil responses can be obtained and communicated to doctors at central sites. With these devices it will be possible to measure the reaction to red and blue light and get information very fast, which helps in making clinical decisions.

An example is shown illustrating how the pupil response increases as a function of light intensity with red and blue light, including the characteristic sustained pupil contraction with bright blue light, which is associated with the intrinsic activation of the melanopsin retinal ganglion cells.

Photo-Electromyogram

My last topic, the photo-electromyogram (photo-EMG), is a very, interesting reflex that has important clinical applications. If one is subjected to bright light, an involuntary blink reflex occurs. It is likely a built in protective reflex for the eye. In bright light conditions, besides the pupil getting smaller, which acts to limit the amount of light received by the retina, an involuntary reflex causes squinting and blinking. That is another accessory means of limiting the amount of light getting into the eye. If one records from the surface of the squinting and blinking muscles with skin "patch" electrodes, the EMG of the orbicularis and squinting muscles will show activation by light, even before a blink occurs.

We wanted to investigate the photic-EMG reflex in more detail as a means of diagnosing abnormal sensitivity to light, which can occur with a number of pathologic conditions of the eye and brain, including migraine and traumatic brain injury (TBI). Similar to the corneal blink reflex, the photo blink reflex is also mediated by the trigeminal sensory nucleus in the brainstem. Some theories propose that in patients with pathologic light sensitivity (photodynia or photo-allodynia), the gain control of the sensory trigeminal nucleus is set to a higher level, rendering the sensory input more responsive. These patients experience much more discomfort in response to light than normal subjects, but up until now there was no good way of measuring this sensory response to light more objectively.

Patients with migraine experience light sensitivity, even between headache periods. Therefore, one useful application of the photic-EMG would be to objectively characterize migraine and its response to treatment. Anyone who has a migraine knows that it is not easy to accurately convey their degree of light sensitivity and discomfort either during a migraine event or in between events. Patients are often treated with preventative, prophylactic medicine and are asked to subjectively evaluate its effectiveness after a period of time, on the order of months. The evaluation of treatment efficacy is all very subjective and not easy to use as the basis for treatment dose and class of medicine best suited for an individual patient.

The clinical application of the photo-EMG would be to take advantage of it as an objective reflex revealing the central trigeminal sensitivity to light in conditions where this appears to be pathologically exaggerated, such a migraine, TBI, and ocular disorders associated with light sensitivity such as photoreceptor disease and some forms of uveitis. One would be able to objectively ascertain whether someone was photophobic due to pathological causes because they would have an exaggerated EMG response of the eyelids and squinting muscles to light levels that most of us do not have. In migrainers, the photic-EMG response could be used during the time intervals between headaches to aid in the diagnosis and classification of migraine, as well as its response to individual treatment.

The example shown depicts the increasing EMG activation of the orbicularis and procerus/corrugator muscles to increasing intensity of red and blue lights, similar to what was shown for the pupil light reflex. The EMG response is shown as the root mean of the electrical activation at increasing light levels from red light recorded from the blinking and squinting muscles.

Note the EMG response to one-second duration blue light. Similar to the pupil response there appears to also be a sustained EMG response, which may reflect activation of melanopsin and may imply that the melanopsin containing retinal ganglion cells also participate in this reflex. The red light stimulus gives a shorter duration EMG response that is proportional to light intensity. The blue light response does too, but as indicated above, it is much more prolonged. The input to the trigeminal nucleus likely derived from the melanopsin-containing retinal ganglion cells whether they are activated by rods, cones or intrinsically via melanopsin. If this is the case, one would also expect to observe a photic-EMG even in patients with photoreceptor disease lacking in rods and cones, similar to what is observed with the pupil light reflex.

If the pupil contraction amplitude is plotted as a function of log stimulus intensity, a sigmoidal shaped response curve is obtained. Similarly, when the maximum EMG response is plotted in the same way, a sigmoidal response function is also obtained, but the half maximum response is reached at a brighter light than the pupil light reflex and this may reflect that the squinting and blinking light reflex may be more of an accessory means to supplement in the reduction in pupil size to limit retinal luminance. However, our expectation is that someone who is photophobic would produce a stimulus-response function of the EMG shifted to the left, revealing a much greater sensitivity to light.

At present, we are right in our infancy with this test. We think it has potential to become an objective way of testing sensory trigeminal function in TBI, migraine and photophobia. TBI patients also frequently get post-traumatic migraine, which is difficult to assess by subjective means.

The photic EMG may provide an objective vehicle upon which to assess these patients and their response to treatment.

Conclusion

In summary, we are trying to use some of the natural rudimentary brainstem reflexes to give us information on patients who are unable to give us reliable information directly because they may have cognitive impairment. This is a very high impact area of research; one that can be translated to clinical practice with development of low cost instrumentation that would be extremely easy to use.

Thank you again to the TATRC division of the DOD for support of this research and to my esteem colleagues in the audience that have provided helpful input and advice.

Discussion

>> ARTHUR JAMPOLSKY:

Randy, how do you separate the eyelid as being the sole recipient of the light response?

>> RANDY KARDON:

We're using the eyelid as the efferent part of the light reflex. You are probably well-acquainted with the corneal blink reflex, in which the trigeminal sensory nerves get stimulated by a stimulus on the cornea. That signal gets conveyed to the brain so that there is a reflex that stimulates the facial nerve causing an involuntary eye blink. The photo blink reflex is similar but instead of stimulating the cornea, a light stimulus activates the recipient sensory trigeminal nerves in the brainstem. Thus, we are using the EMG of the blinking and squinting muscles as an output to tell us how much light was sensed by the primitive brainstem centers of the brain. So we could also use this reflex as something similar to the pupil light reflex, but for measuring afferent transduction of light from the eye similar to the pupil light reflex. We also need to determine how the photic EMG is affected in the unconscious state, such as anesthesia, because the reflex may still be measurable, providing a way of assessing visual function in the unconscious state.

>> AUDIENCE MEMBER:

So how do you distinguish between the reflex response to light and modulation by the cortex? For example, we know that the pupils can dilate because of surprise. Obviously there are two components that can influence brainstem reflexes; there is a direct light input component and a supranuclear modulating component. It would be desirable to separate them somehow.

>> RANDY KARDON:

We know there are a number of supranuclear pathways that influence the pupillary reaction besides afferent input. If I give everybody in this room the same conditions of a light stimulus, we would get quite a large variation on how much the pupil responds partly due to variations in retinal sensitivity, but also partly due to the supranuclear state of the person, i.e., how excited or drowsy they are. Part of the beauty of using the red and blue light is to attempt to internally normalize the blue light contraction amplitude to the red light contraction amplitude under matching retinal illuminating conditions and then assess the amount of sustained pupil response in relation to the initial contraction amplitude. This may take into account the status of a person's supranuclear inhibition to the pupillomotor system.

>> RONALD SCHUCHARD:

Randy, I am wondering about the use of eye movements to assess visual input. All of the work I know has shown that using eye movements as a response by subjects, participants, and patients, is usually a huge problem. It does not correlate to visual perceptual thresholds.

>> RANDY KARDON:

Using eye movement perimetry.

>> RONALD SCHUCHARD:

By eye movement.

>> RANDY KARDON:

The way we are planning on doing this is by showing a target to a subject moving across the screen that one attends to. The subject's eye position will correlate to where the target is located in visual space. Then, as the target is moved, the attributes of the target, such as contrast or spatial frequency, are changed and when that visual target drops below the threshold of perception, the eye position will no longer correlate with target position.

>> RONALD SCHUCHARD:

Sure that extreme is obviously true. My concern is before you get to that extreme. They can lose eye movement correlation to your target long before they lose perception.

>> RANDY KARDON:

Instead of becoming less clear or less formed, the vanishing optotype by its construction will disappear on the background when it drops below threshold. Its appearance is binary; it either appears or disappears. So the idea is to tweak that optotype so it will correlate with a measure of standard measure of either low or high contrast visual acuity. In addition, newer eye and head tracking technology and novel signal processing approaches allow one to minimize noise and the variability that I do not think was possible in the past.

>> RONALD SCHUCHARD:

The technology is not the issue.

>> ARTHUR JAMPOLSKY:

We tried something like this years ago with moving contrast optotypes and tracking eye movements and if you started at high contrast where the subject is really good at detecting the target, then the correlation between eye movement and target position falls off almost immediately and is related to visual contrast sensitivity. I do not perceive a major problem if your target is moving at a speed of approximately 5 degrees per second. We had a target moving and every couple of seconds the target contrast decreased as it moved randomly across the screen. There is a pretty discrete point of loss of eye tracking related to visual threshold perception. I am less worried about it than the other audience member.

>> RANDY KARDON:

I was trying to respond to Ron's question. I think it is a task-dependent as far as the eye movements are concerned. A good task is finding something with your eye movement. This will show a correlation, some process that is potentially of interest to you. The criterion for finding something is, of course, the combination of recognizing, otherwise you cannot tell whether somebody found it or not. You do not even have to measure the eye movement. If someone finds a target somewhere, let's say to a red square or whatever, and is unable to identify it, to say it is red or a square or whatever, then you have an indication that the person found it. If you measure the latency of that eye movement you get around the technology question as to how to measure eye movements. You still you have an indicator of that recognition process.

>> WILLIAM GOOD:

Now, it is my pleasure to introduce Michael Gorin. Michael holds the Harold and Pauline Price Chair in Ophthalmology – and is a professor of ophthalmology.

Light Sensitivity in mTBI:
Models for Assessment and Causality

MICHAEL GORIN, MD

I am a newbie here because I am probably the least experienced in actually studying traumatic brain injury. Nonetheless, I have been approaching this field through my work.

I am actually a molecular geneticist and ophthalmology geneticist. My last visit to Smith-Kettlewell was 30 years ago. Smith-Kettlewell first funded my family-based genetic studies that led to the eventual discovery of the first major genetic loci and genes that contribute to the risk for age-related macular degeneration.

I am deeply appreciative of this opportunity to join others in this field. Basically, what happened was that I got involved in the study of photophobia (light sensitivity) because I was intrigued that women carriers of X-linked cone dystrophy were readily identifiable due to their light sensitivity. That unexplained observation set me on a nearly 20 year path to understand the etiology of light-related pain and finally to its relationship with traumatic brain injury. I am not even going to talk about my work on pupillography, which was started for completely different reasons. That is a project in which I have been trying to figure out a way to non-invasively assess retinal ischemia, a major problem in diabetics, using the pupillary response as a biosensor. However, as I have learned more about the biology of the pupillary pathway, I have realized that this work has also converged with my interest in the impact of traumatic brain injury on light-related neural pathways.

Light-Associated Allodynia

In that respect, Randy and I share a lot of interest in pupillography. We switched to using the term "allodynia" because photophobia has connotations of being fearful. When I was investigating a large family afflicted with X-linked cone dystrophy, I was impressed as I went from farmhouse to farmhouse, examining family members. I could pick out the women who are carriers, because they were all light-aversive under normal lighting conditions. Individuals affected with hereditary cone disorders were often so light-sensitive that they were more disabled by this condition than by their actually visual impairment. I also saw patients in my regular clinic with idiopathic photoallodynia who were literally disabled by their light aversion, even with normal visual acuity and the absence of ocular pathology.

As I got into the study, I became aware of how big an issue photoallodynia is for our Veterans who have traumatic brain injury. Once I started out looking into this, I came to appreciate the need for better understanding in this area and I am endeavoring to acquire the new skills and tools to pursue this avenue of research. As I am starting to pursue this new phase of my research career, I will hopefully learn a lot from this meeting.

Light-associated allodynia (LAA) is a condition in which pain arises from exposure to normal levels of light. While hyperalgesia is an increased sensitivity to pain, which may be caused by damage or activation of nociceptors or peripheral nerves, allodynia represents the perception of pain that is mediated through a sensory pathway that is not intrinsically for pain. Light aversion to a sudden exposure to light is considered a dazzle response and differs from light-associated allodynia that is a sustained painful response. The difference is that dazzle and even normal light aversion responses are not necessarily painful. Analgesics, more commonly known as painkillers, are used to reduce the feeling of pain but often have little or no effect for light-associated allodynia, again suggesting that pathways other than the traditional nocioceptive pathways may be involved.

In contrast to light-associated allodynia, "glare" occurs when you get loss of retinal contrast due to stray light. It is important to appreciate that this is not considered a pain-related response generally, and must be considered as a separate symptom in the patient population about whom we are talking. This list shows reported light sensitivity, we have seen other estimates. Also on the list are headaches and sleep disturbances, categories I will come back to later.

Obviously, this is not a trivial issue and it has been brought up before. As I said, I got into light-associated allodynia because of cone dystrophy. I was looking at it from a peripheral point of view (focusing on the eye and the retina as the sensory organ for light-related pain) and recognizing that, yes; you can have an injury to the eye that can be the primary cause of light-associated allodynia. Then I started to realize that there are reports of brain lesions that lead to photophobia; pontine lesions, and other migraine and traumatic brain injury, being the major ones. What is the interrelationship of these conditions? Our only therapy involves giving these people darkly-tinted lenses. Yellow filters that filter out blue wavelengths seem to be more effective than giving them gray tints. I think it has to do with the spectral sensitivity that Randy alluded to earlier.

Genetic Approaches

First of all, as I got started I decided I needed a behavioral assay for a mouse-based model that could reliably separate out light aversion behavior from anxiety. By the way, despite their preference for nocturnal activity, the mice are not "naturally" photophobic. We wanted a mouse model for light-associated allodynia and light aversion because of our ability to selectively eliminate specific cellular functions and cell types using transgenic animals. There are well-characterized mouse mutants for which rod photoreceptor function and/or cone photoreceptor function can be eliminated. There are other opsins in the eye in the pigment epithelium, and we have used knockouts of those as well. We can eliminate all of the functional rod and cone opsins using animals lacking RP65 activity or selectively eliminate rod or cone function using GNAT1 or GNAT2 knockout animals. We have also used diphtheria toxin under the control of a melanopsin promoter. When an animal matures, as it tries to express its melanopsin gene, it releases diphtheria toxin. We have ablated the entire population of ipRGC retinal ganglion cells without affecting the rest of the ganglion cell layer of the mice. And now we have acquired mice that have the same method of cell ablation using a cone photoreceptor-specific gene promoter so that we can test the difference between a retina that has inactive cones (such as due to a GNAT2 mutation) as compared to a retina in which the cone photoreceptor cells have been physically eliminated.

Our assay is really pretty difficult. You have a light/dark box comprised of two chambers and a variable light illumination. We monitor the animal's behavior in these chambers under varying conditions using infrared imaging of the mouse movements. This light/dark assay was originally developed as a measure of mouse anxiety. If mice are anxious, they tend to stay in the dark. Thus we have gone through testing and training periods to establish that the mice are very relaxed and calm in the test environment. You really have to have use atropine, particularly when pupillary responses may be altered by either the genetic or pharmacologic manipulations. Thus we routinely have all of our animals fully dilated with atropine before undergoing testing. When you are comparing an albino mouse versus a pigmented mouse, you are not comparing the same thing if the extent of dilation is not adequately controlled.

Light Aversion Therapy

We recognize that we are actually measuring a light aversion behavior that may or may not be associated with the perception of pain. We have used a variety of analgesic and anti-inflammatory agents to establish the nature of the light aversion behavior and found almost no effect on the levels of light aversion behavior in all of our mice until we used low-dose morphine, (below the dose that causes hyperactivity). Consistently, in every test condition, this drug elicits a severe light aversion response. We were very startled to see this.

When I discussed this finding with one of my UCLA colleagues who is an expert in migraine disorders, his response was: "Of course, you should never give opiates to migraine patients. It makes their light sensitivity worse. Those of us who actually deal with these patients know that the opiates are really bad for these patients." We found that, unless you completely wipe out phototransduction, you can see light aversion behavior even in mice that are completely lacking rod and cone function. Only when we also eliminated the ipRGCs do we see a decrease in light aversion behavior. You do not even see a pupillary response in these animals and yet their response to the low-dose morphine persists.

Mechanisms of Light-Associated Allodynia

We confirmed that when we reach that level of ablation of the ipRGCs that reduces light aversion and the pupillary responses have been eliminated, the mice's circadian rhythms are thrown off. They cannot adapt if you shift the timing of the light/dark intervals. We now see that there is a common pathway shared by light-associated allodynia, circadian disturbances, and pupillary responses. Interestingly enough, the morphine-reduced response is not completely ablated even if you knock out all of the ipRGCs. We do know that the effect is the due to the actions of morphine on the mu opiate receptor and is reversible with naloxone.

We think that the morphine-induced light-associated allodynia is most likely a central mechanism, which brings it very much into the arena of the discussion of today. The ipRGCs seem to be the primary mediators of this non-image forming visual pathway. By the way, ipRGCs use a phototransduction pathway that is more similar to invertebrates than vertebrates and seems to be an evolutionarily very ancient vision system. In the mammalian eye, you're actually observing a more sophisticated vision pathway for formed image perception that is layered on top of this ancient pathway that is used for more fundamental light-related biological processes. This ipRGC pathway relies on the neuronal firing triggered by light activation of melanopsin and it takes bright light of 480-85 nm, which results in continuous neuronal firing as opposed to the short bursts of neuronal activity caused by rod or cone photoreceptor activation, explaining how continuous light can cause continued pain.

We have a very simple model that normal light is activating photoreceptors or ipRGCs directly. Then there is a shared visual pathway that splits off in the brain to control the pupillary response, another for the light avoidance in the brain, and another for the circadian entrapment. You may have lesions that affect common components of all three systems, light avoidance, circadian photoentrainment and pupillary responses, or maybe just some of them.

Relation to mTBI

One of the questions is going to be: What is the correlation of the three? I think this is a really important question. We are predicting that partial associations with light are probable. I do not have access to the data on humans that is available to those who work for the DoD and VA that asks whether there are cross-relations. The most common central cause of light-associated allodynia is migraines. It is on the order of 70-80% of mTBI patients who actually have post trauma migraines. So we actually believe that they are related.

You see 81% of the participants in this study of patients with post-concussion and chronic head injuries had headaches. Sixty-five percent had light sensitivity. How many of these are individuals who have combinations of these symptoms, perhaps reflecting damage in shared brain regions? This raises additional questions. If there is a consistent relationship between migraines and light-associated allodynia, then we have potential target strategies for therapies.

Calcitonin gene-related peptide (CGRP) is known as key factor in migraine pathogenesis. There are now molecules available to block this. There are also other molecules that are very much related to CGRP, which are potential therapeutic targets. This combination raises possibilities that we may be able to try to use some of these agents for patients who are experiencing light-associated allodynia even if they do not have classic symptoms of migraine headaches. Perhaps it would not be a bad idea to do a clinical trial with one or more of these drugs to see if there are benefits for patients who are severely photophobic but do not have migraines.

I am a migrainer. As a geneticist for over 20 years, I wondered from whom in my family I got it? One of my sons and my brother had migraines and yet my parents denied any history of headaches. Only late in life, did my father reveal that he had a history of visual migraines (the light flashes reported by many patients) but without headaches. It is certainly conceivable to think a person could have photophobia through a central mechanism that is shared with the etiology of migraines.

There are a number of challenges as a researcher. First of all, we need to know whether or not the light-associated allodynia induced by low-dose morphine is related to the central brain disturbances that are caused by the head injury associated with mild TBI. How does brain injury alter neurotransmitter levels and receptors in selected regions of the brain that might alter a patient's response to certain environmental, light and drug exposures? Using the genetically-engineered mouse in conjunction with pharmacologic manipulation as well as models for creating mTBI we can begin to unravel the relationships of light-associated allodynia with pupillary and circadian disturbances. From the perspective that I have tried to present in this discussion, it may be worthwhile to treat a circadian abnormality or sleep disturbance in a patient who also has photophobia and we may be able to elicit a favorable response. Many things we could do clinically I think are amenable to this "holistic" approach.

As I said, we found out just recently when we use animals whose cone photoreceptors are blind (but present) as compared to animals whose cone photoreceptors are selectively ablated by diphtheria toxin, there is a difference in their light aversion behavior. We suspect that there is a connection between the rod photoreceptors and the cone photoreceptors, which mediates signals in the absence of cone-related light responses that are passed to the bipolar cells and ganglion cells connections. While we believe that light-associated allodynia, or photophobia, associated with mTBI is generated by pathology in the central brain, there are unique aspects of retinal circuitry that may play a role in initiating the light-dependent pathway and may be modulated pharmacologically to reduce symptoms.

We can manipulate migraines and perhaps reduce symptoms in patients. Am I going to try it in any of my patients? Probably. Some are truly, truly desperate. Their severe light sensitivity is not just disabling, it is socially isolating. For some of these individuals, anything we can do is worth trying.

I did recently have a patient who was on narcotics due to chronic pain syndrome. Based on our studies, I weaned him off the narcotics and his symptoms have seemed to have improved. Our patients can be an invaluable source of new insights and opportunities to better understand the nature of this condition.

I am just starting out here. I am hopeful that we will be making some inroads so we can, not only help patients with TBI, but all of those who suffer from photophobia. I have lots of other stuff I could talk about, but I am not going to overextend my time, so thank you.

Discussion

>> WILLIAM GOOD:

Fantastic. Thank you. Questions?

>> AUDIENCE MEMBER:

At times you hear about some alleviation in the sensitivity using differently colored tinted lenses.

>> MICHAEL GORIN:

I am an advocate for blocking blue light in patients. If you use gray, you are essentially reducing the overall illumination of the retina, which can impair vision with no greater reduction in the perception of pain. If you filter out just the blue light, you then have ample amounts of light and you can then see effectively. You also do not trigger even greater dilation. I do not know about other clinicians, but I only do retinal exams with yellow-tinted lenses. Unfortunately you cannot get the examination lenses made with yellow glass anymore. It makes a huge difference for even my non-light-sensitive patients by allowing them to more comfortably cooperate with the retinal examination.

>> AUDIENCE MEMBER:

There are some sunglasses on the web called Cocoons with orange-looking lenses. I have been using them on a number of patients who seem to really respond.

>> MICHAEL GORIN:

I think there is a combination, just as Randy alluded to, of different tints that can work well for a variety of patients. When I started in this field from the perspective of a retinal geneticist, I had the hypothesis that both rods and cones play competing roles. My simple-minded model used to be that the stimulation of the rods tended to be stimulatory of a photophobic response (since those photoreceptors are more sensitive to light and perhaps more prone to light damage) and the cones responses were inhibitory. When patients have a cone dystrophy the functional loss of the cones creates an imbalance in the input to the light-associated allodynia pathway leading to an increase in pain. When the rod photoreceptors are eventually lost, then the light sensitivity would abate. When rods and cones are both degenerating, it seemed that the degree of light sensitivity was dependent on the relative balance of the remaining rods and cones. We have not found such a regulatory system in the mouse, but it is also clear that mice have retinas that are dominated by rods and the cones play a much lesser role than they do for primates. In addition, I now have a much greater appreciation for the potential role of the ipRGCs, which can detect light independently of the rod and cone photoreceptors.

>> CHRISTOPHER TYLER:

Two points. One is when I put the yellow lenses on, I immediately feel happy, as though the sun came out - both the rose-colored and the yellow-colored versions - a striking phenomenon for me. And the other is that the perceived contrast is notably enhanced by filtering out the blue light.

>> MICHAEL GORIN:

I think a lot of people have a certain degree of discomfort from blue light, especially in supermarkets. If you ask which I recommend, when I was in Pittsburgh, I told everyone to go to hunting stores. Everyone knew where to find the hunting stores. In California, I tell them to go to ski shops.

It is the same kind of lenses that filter out blue, especially with side wrapping. The blocking of the light from the sides of the glasses makes a huge amount of difference. Part of the enhancement of contrast is that by filtering out the blue light before it enters the eye, there is reduced light scattering caused by the yellow fluorophores that are accumulating in the natural lens as we age.

>> AUDIENCE MEMBER:

Pilot studies on yellow lenses also say, if given the opportunity, always use yellow. On the other hand, people who are skiers also – with very good control studies – know the snow is not white. It has a lot of blue in it, as any skier knows.

>> *AUDIENCE MEMBER:*

Mike, will you define "dazzle" for me?

>> *MICHAEL GORIN:*

Dazzle is the very immediate response to a single flash of light, like photo-blink response. It is that momentary response. Also, there is an autosomal dominant condition of sneezing upon exposure to sunlight or a bright flash. It is a dominant condition (the gene involved has not been identified), so the risk of transmission to each of one's children is 50%. Dazzle is not a painful response, just a physiologic response to exposure to a sudden bright light.

>> *AUDIENCE MEMBER:*

How short a flash of light can cause a dazzle response?

>> *MICHAEL GORIN:*

Generally, I would say on the order of less than 200 milliseconds. It can be as short as 10 to 50 milliseconds.

>> *AUDIENCE MEMBER:*

Will you perceive it?

>> *MICHAEL GORIN:*

You will perceive it, and you may avert or close your eyes, but it is not sustained. It is much like a person's response to the flash from a camera. If it were a sustained exposure to light of that intensity they would be likely to experience a light-induced aversion response, which might be associated with discomfort. This would not be light-associated allodynia since this is at a level of brightness that would normally trigger an aversion response.

>> *AUDIENCE MEMBER:*

Is it a relationship of the retinal image?

>> *MICHAEL GORIN:*

Not that I know of.

>> *AUDIENCE MEMBER:*

Could the exposure to a focal intense flash of light or laser cause pain through a mechanism that is distinct from light-associated allodynia?

>> *MICHAEL GORIN:*

I understand. Perhaps. I am just trying to think of the difference between a focused laser light versus the other. We have not talked about why you get pain from laser burns – that is another issue.

>> *WILLIAM GOOD:*

Isn't there a stroke syndrome, dazzle -- you get a dazzle effect?

>> *AUDIENCE MEMBER:*

Second question: Talking about opiate receptors, opium takers have very small pupils.

>> *MICHAEL GORIN:*

Paradoxically, mice pupils react to morphine with dilation.

>> **AUDIENCE MEMBER:**

Should I think of that as a photophobic response?

>> *AUDIENCE MEMBER:*

No, it is central. It is the Edinger-Westphal nucleus. Normally you have these inhibitory fibers coming from the reticular formation that keep your pupils at midsize. The neurons in the Edinger-Westphal nucleus are very unusual. If you see, you afferent them, and they fire like crazy all the time. When you start to go to sleep that inhibition goes away and your pupils get small. That is why narcotics in humans make the pupils get small and why under anesthesia the pupils dilate.

>>*WILLIAM GOOD:*

Greg Goodrich's topic is 'Self-assessments and Their Effectiveness: What Vision Tests are Currently Used?'

Self-Assessments and Their Effectiveness:
What Vision Tests are Currently Used?

GREGORY GOODRICH, Ph.D.

Self-Report Tests

John and I are looking basically at self-reports. For the Monty Python fans in the crowd; now for something entirely different; instead of nicely organized, laid out research, we're going to talk about policy clinical research.

Basically what I want to talk about is the Rivermead Behavioral Memory Test being referenced. There are a wide variety of self-report questionnaires that are used in vision. The question is how you can use the quality of life for whatever the study is? We looked at 161 consecutive patients coming through the polytrauma rehabilitation center. These were all patients diagnosed with mTBI who came into a clinic for one of the variety of things: chronic pain, general medicine, hearing, vision, neuropsychology, and so on. We have a binocular vision screening that we do with these patients.

We were interested in what we could do as a self-report that might help us better assess these patients. Our starting point was basically the adaptive visual symptoms questionnaire on the second page of the handout. We adapted it and started using it as a yes/no question. We just asked the patient sitting in the waiting room to check off yes/no on symptoms.

Patient characteristics were as follows: the median age was 28. They are very close to the age population that, again, was talked about earlier although we did pick up some people from the Vietnam and Korean War in this population. I am not sure how they got into the study, but they were included. Most of the patients we saw, 72-73 percent had mTBI associated with a blast event. The remainder had TBI due to motor vehicle accidents, falls and other causes. One interesting study now is to look at blast injuries and vision loss compared to other trauma injuries all in current or recently discharged troops.

Effects of Combat vs non-Combat Injury

The study that preceded this one looked at combat versus noncombat injuries. The interesting thing was that they were not more likely to have a gunshot wound in the non-combat than in the combat group. Perhaps it may be a commentary on society, I don't know.

This population, again, is normal visual acuity, normal fields, only 2 percent had visual acuity such that they were legally blind; and less than 3 percent had a significant visual field defect. They are all presenting themselves as visually normal in a generic sense. We asked about their vision related complaints following injury. Virtually none of these patients were coming into the clinic to see the optometrist. They were coming in for other reasons. We found high rate of binocular dysfunction associated with the polytrauma brain injury population.

The mild TBI patients reported a visual related complaint. When we asked specifically about reading, 88 percent said that they have a reading problem. There are a lot of reasons you can have reading problems. This list just provides you the percentage breakdowns by visual complaint.

We did find some relationships between the questions asked with this dataset and our binocular clinical findings, although it was surprising how low the correlations were. Some were statistically significant. Two self-reports were associated with an accommodative insufficiency and pain in and around the eyes and pain with history of refractive error. I am not sure what this indicates. We found it, so I'm reporting it.

A history of refractive error was also associated with convergence insufficiency. The patients were unable to do sustained reading because the eyes get tired of reading, pursuits, saccades deficit, when resting were associated with fixation dysfunction. These are significant relationships, but I think we need to take them with a grain of salt partly because we were asking yes/no questions. It is not a graded scale. Unfortunately, I do not have the data analysis to report to you on differences. But my take-home message from asking patients to fill out a visual symptom checklist is that it provides the clinician with some useful information. It will not tell you much about who is going to have an accommodation insufficiency or convergence deficit. So there is general information there about self-reported symptoms.

The other thing is that I think this information is helpful is in trying to sort out the affects of visual dysfunctions, PTSD, depression, and other sorts of things. If we can compare similar system checklists, are there similarities and differences there that we can find between those that will help us sort out the different visual conditions?

There is a lot more that could be done in this topic area. I do not have the knowledge of this to share with you yet.

Discussion

>> AUDIENCE MEMBER:

One of my colleagues a number of years ago had the hypothesis that patients who were depressed would describe dimness in their vision and he started looking at that. Does the question of a patient's perception of dimness, ever come up in the questionnaires? Is it correlated with any other psychology assessment?

>> GREGORY GOODRICH:

Not to my knowledge. I have also heard reports that depressed people feel a constriction of the visual field. Is that the same as dimming or not? I'd say 'yes'.

>> AUDIENCE MEMBER:

Pursuits, saccades, deficits linked with renal problems. Is it more of the saccadic system and more of the pursuits?

>> GREGORY GOODRICH:

We have not tried to sort it out. Anybody want to take a shot at that?

>> WILLIAM GOOD:

Steve, can you elaborate on that?

>> STEPHEN HEINEN:

Actually, I'm not sure even if anybody has done the first part. We know that there are pursuits and, separately, saccade issues with TBI. We know that there are reading issues with TBI. Has anybody tried to put those two together? Are they correlated?

>> AUDIENCE MEMBER:

They can be correlated because of existing in the same impact event.

>> KENNETH CIUFFREDA:

I will try to answer some of this tomorrow. The short answer is yes.

>> WILLIAM GOOD:

I have a practice full of kids with nystagmus. I will be interested in hearing your presentation.

>> *AUDIENCE MEMBER:*

If you test the reading rate, they probably do not read okay. They read slower, but they read.

>> *AUDIENCE MEMBER:*

I noticed that a lot of the questions in the questionnaire center on the general phenomenon of reading. I say "general phenomenon", because it is so messy. In reading there is a feedback cycle between the reading material and the reading person.

Do you know of any study that has tried to manipulate this experimentally by changing the reading material? It matters what you are actually reading. For instance, a text that makes some sense is different than for words that do not connect to anything. There are many possibilities in that regard. Has any work been done in this area?

>> *AUDIENCE MEMBER:*

The questionnaire is not actually measuring reading.

>> *GREGORY GOODRICH:*

I am not so sure.

>> *AUDIENCE MEMBER:*

You know there is a point of not being able to remember what you just read. You can influence the degree of difficulty someone might experience by manipulating the degree of difficulty of the text they have to read.

>> *GREGORY GOODRICH:*

Yeah, I think there are a lot of fruitful areas. We were doing self-report questionnaires to see how it might inform our clinical assessment.

>> *AUDIENCE MEMBER:*

Have you had the generation of patients complain of difficulty gaming?

>> *GREGORY GOODRICH:*

No, and we have not asked about it. But one of the things we are bringing in to our center for more severe vision loss is a weak gaming symptom. I do not recall any reports on this.

75

>> GABRIELLE SAUNDERS:

Certainly we know that hearing aid outcome is influenced by many non-audiological factors, such as motivation, expectations, spousal support and psychological wellbeing. It is also interesting that often patients have not thought much about their hearing problems when they come to the clinic/laboratory. We often find that that having them complete hearing questionnaires opens their eyes to the hearing difficulties that they are having, and to their needs and desired hearing rehabilitation outcomes.

>> GREGORY GOODRICH:

That is a salient comment. Our clinicians think the patients come in better prepared for the exam.

>> GABRIELLE SAUNDERS:

I suppose the big question is, does this also make them more open to the intervention and rehabilitation?

>> GREGORY GOODRICH:

Yes, although I have to say we diagnose a fairly high percentage and ask if they want to be seen in the general optometry clinic or make an appointment for binocular therapy. Often the answer to the latter is no. It is not on their priority list, although some people do.

Some will come back later as they start to recognize it as a problem. But I think if you are dealing with readjustment from a deployment, trying to find or maintain a job, get to school, deal with PTSD depression, substance abuse, whatever, the whole litany of things, it does not seem to be high on their priority list.

>> WILLIAM GOOD:

Now, it is a pleasure to introduce Suzanne Wickum, who is visiting us from the University of Houston. She is an expert in pediatric binocular vision. Her title is 'Oculomotor Function Tests, Photosensitivity: Accommodation and Convergence Testing in mTBI.'

Oculomotor Function Tests, Photosensitivity, Accommodation and Vergence Testing in mTBI

SUZANNE WICKUM, OD

Since most of you do not know me, I want to start with a quick background. I come from the clinical trenches and not so much from a research standpoint. I am seeing brain injury patients of all levels, mild, moderate and severe, both at the University of Houston College of Optometry in an outpatient setting as well as at an acute in-patient setting at The Institute for Rehabilitation and Research (TIRR) at Memorial Hermann Hospital in Houston; one of the top five rehab facilities in the U.S.

You have probably heard about TIRR Memorial Hermann in the news recently; Gabby Giffords is there now. This afternoon, I speak from the clinician's perspective and will go over what we do in all of these areas. It is a lot to cover in ten minutes.

One earlier question was about what tests are used to look at binocular vision and accommodation in the clinical realm. I will provide some information for you.

Even though I am not in a VA setting, most of the mild brain injury patients I work with happen to be military. TIRR Memorial Hermann has an affiliation called *Project Victory*. Soldiers come to us from all over the country to undergo comprehensive brain injury rehab. TIRR sends us all of the Project Victory soldiers who have vision issues.

Symptomatology

The soldiers we work with seem to be a different group than what I like to call our "garden-variety" brain injury patients. The combat-related and the blast-injury related brain injury patients seem to be a little different than non-blast brain injury patients.

A symptom survey (the convergence insufficiency symptom survey, CISS) is part of our workup. When we have looked at symptoms, our survey tends to show that some type of binocular and/or accommodative problem exists and it may or may not be convergence insufficiency. The binocular vision and/or accommodative dysfunction could be anything but the CISS helps us target our exam testing for the patient. We did a retrospective review of 26 blast related TBI subjects and evaluated their symptoms and

77

diagnoses. One of the top symptoms we see in this population is photophobia, or light sensitivity, which was present in 85% of our subjects. Other common symptoms reported were: headaches (81%), reading difficulty (50%), asthenopia (42%), motion sickness (27%), and glare (23%).

With our retrospective study population we had no problems reported with eye-hand coordination. In this population binocular and accommodative disorders predominated. Vertical phorias were the most common binocular vision disorder noted (31%) which contributes to patient symptoms of dizziness (31%) and loss of balance (8%). The incidence of vertical phorias was much higher than the average non-brain injured adult population (0.5%).

Vertical phorias tend to occur much more frequently in mTBI and is a finding that is often missed. Other binocular and accommodative dysfunction commonly found in our study included: accommodative infacility (23%), convergence insufficiency (15%), and accommodative insufficiency (15%), which round out the top four most common entities.

Ocular Motility and Eye Movements

What do we do to diagnose all of this? We have a whole bunch of tests clinically that we look at and I will run through them with you. Extraocular motility testing helps diagnose cranial nerve III, IV, and VI palsies and that will help guide us in our additional testing throughout the clinical exam.

Questions about reading and eye movements have come up. This is an area that I find is a little bit tough clinically. Without actual eye tracking devices, we do not have great tools for measuring saccadic performance clinically. The average clinician that might be working with these patients likely will not have eye-tracking devices in their office and are thus left to subjectively grade saccadic performance. The Northeastern State University College of Optometry (NSUCO) oculomotor test was devised as a "standard method" of scoring saccadic and pursuit eye movements. It evaluates things like how many undershoots and overshoots are made during saccadic eye movements? Does the patient turn their head or body rather than their eyes to look between targets? Similarly the NSUCO test has a scale for evaluating pursuits. Other scales such as the SCCO scale for saccades and pursuits have also been developed but none of these tests are used nationwide.

Other tests that people have talked about today which look at saccades are the developmental eye movement (DEM) test, which was created to help evaluate reading issues. The King-Devick (KD) test was featured in an article that just came out in *Neurology* that uses number reading as a screening tool for concussion (Galetta et al., 2011).

Let me back up to the DEM test. There are a lot of studies looking at these tests in children with reading dysfunction. Is the child's reading deficit due to a saccade problem? It is kind of chicken-and-egg situation. There is data on both sides. I do not think of the DEM as a purely saccadic eye movement test, because it does involve reading aloud a series of horizontal and vertical numbers thus requiring cognitive processing, number recognition, speech/language skills, etc. The King-Devick test is being looked at as a sideline screening tool for concussive sports brain injuries. The *Neurology* study that came out from the University of Pennsylvania found that pre-injury versus post-injury time worsened by three seconds or more in people who have mild brain injury/concussion. They looked at boxers and mixed martial arts fighters where the patient loses consciousness and evaluated KD scores before and after concussion. The KD test has a series of numbers that the patients have to read as quickly as they can. There is a training run and then three different target pages where the patient has to read across all the numbers as quickly as possible. The University of Pennsylvania study sample size was only 39. They are now looking at larger sample sizes.

Near Point of Convergence

We need to do the *Near Point of Convergence Test* several times in order to look for the fatigue factor. A lot of brain injury patients have issues with fatigue. If you repeat the test and you fatigue the patient's system, then you can diagnose certain conditions more easily. As a doctor, I like to perform this test objectively (watch the patient's eyes for loss of convergence) especially when people doubt a patient's sincerity about their symptoms. When I first started working with soldiers, I wondered if they were faking things so that they could get out of the military. I quickly learned that, if anything, the soldiers withheld symptoms to get back to combat more quickly. Some Workmen's Comp patients do not always have the best motives. Objective tests are better in such cases than subjective tests, which can be very unreliable.

Distance and Near Cover Test

Distance and near cover test provide an objective evaluation of ocular alignment, strabismus, comitancy, and AC/A ratio. During the cover test we are also really looking for vertical deviations and using the cover test for *Park's Three Step*, which is the test that we do to assess for cranial nerve IV palsies.

When suspicious of a vertical deviation based on patient symptoms, but not necessarily seeing the vertical on cover test, I can get the patient's input as to whether there is a vertical misalignment. This card here (Modified Thorington Card) is to help give an accommodative stimulus during Maddox rod testing at near so that the patient has a similar response to their cover test where you are using a near accommodative target. There are small numbers and letters on the card, which represent prism diopter values so that the magnitude of the subjective deviation can be measured.

Associated Phoria Testing

I use *Associated Phoria* testing a lot to help prescribe prism. Brain injury patients often have so many other things going on in their life, they are not able to tolerate or add vision therapy into their daily schedules. Prism in glasses can often help alleviate symptoms from BV disorders. If the patient then wants vision therapy in the future, so that they do not have to use their prism glasses, we can address that as their life calms down again. The Wesson card and/or vectographic slide help us figure out what the patient's associated phoria magnitude is and that value is typically used as the starting point for prism prescriptions.

Vergence Testing

We frequently assess positive and negative fusional vergence ranges in free space with prism bars. In addition, to watch for patient fatigue we may test vergence facility. Normal vergence facility is 15 cycles per minute using 3 pd base-in to 12 pd base-out prism jumps. As we are flipping the prism back and forth, stimulating negative and positive vergence, we look to see if the patient fatigues.

Sensory Testing

We look at both global and local stereopsis. If the patient does not demonstrate stereopsis (third degree fusion), then we want to know, if they have second-degree flat fusion. We use the *Worth Four-Dot Test* to determine if the patient has second degree fusion, suppression, and/or diplopia.

Torsion Testing

Any time a vertical misalignment (phoria or strabismus) is present, I measure for cyclotorsion because, if a fourth nerve palsy is involved, there will likely be torsion present which can prevent fusion with prism. Torsion is not correctable with prism. If the torsion is significant, then surgery can be done to help the patient alleviate their symptoms. You can imagine if your world was breaking apart in a criss-cross fashion all the time, it would be quite disturbing.

Accommodative Testing

I do accommodative amplitude (AA) testing with a grain of salt. First of all, this is a subjective test so we need to have a reliable patient response. Brain injury patients tend to have slower responses. So, if you are moving the target for a pushup amplitude too quickly, the AA may look much better than is truly is. And, if you perform a pull-away AA, the AA may look much worse than it truly is. There are also problems with target relative magnification as the target moves closer to the patient. Some doctors choose to perform minus lens AA assessment, which can be contaminated by relative minification of the target.

Dr. Ciuffreda and his group have been studying accommodation in brain injury patients. They have found a decrease in velocity amplitudes and inconsistency in patient responses over time compared to normals.

Along with our accommodative amplitude testing, I like to look at MEM retinoscopy, which objectively assesses accommodative accuracy. This test shows us if the patient has a lead or a lag of accommodation. You can also see if there is a lot of fluctuation in the accommodative response, which I often find with these patients. We also like to assess monocular and binocular accommodative facility. This can help look for fatigue as well as whether the patient has more trouble relaxing or stimulating accommodation.

Photosensitivity and Selective Wavelength Filters

Eighty-five percent of the blast-injured mTBI soldiers told us they had light sensitivity. It is extreme and life-altering photophobia. Their children cannot be in the living room with the lights on because it hurts Dad's eyes.

The selective wavelength filters that we use eliminate the blue end of the light spectrum, some into the blue-green spectrum. We also use polarized filters for overall light reduction and to reduce glare symptoms. We had the soldiers give us a subjective response as they looked through the different filter tints and tell us whether they felt more comfortable or not. The selective wavelength filters we use are 450, 511, 527, and 550 nm as well as polarized filters. The patients usually tell us quite a bit through facial expressions as they look through the various filters. When the patient gets the filter tint that alleviates their photophobia symptoms the most, their posture relaxes and many of them actually look directly into the light. We also add a polarized shield/clip-on over the selective wavelength filters for further reducing bright outdoor light. We can tint lenses in our University optical or send the spectacle lenses out to Chadwick optical for the selective wavelength tinting.

When I first started working with these filters with the photophobic, mTBI population, it did not sit well in my stomach without some background information as to why the filters were working or what the filters were doing. This is very subjective and there is not much data to guide us.

We looked at color vision, with and without the selective wavelength filters, because we knew we were filtering out the blue wavelengths so we documented that the filter induced tritan defects. We found generally equal or minimal improvement in contrast sensitivity with the selective wavelength filters in place. We found the Mars contrast sensitivity chart was a good contrast sensitivity test, much like the Pelli-Robson test.

We also looked at brightness acuity testing due to the patient complaints of glare. The patient's acuity remained good with all 3 brightness levels. The majority had 20/20 or 20/25 acuity and remained so at all levels with and without filters. However, the body posture became more tense and the discomfort level increased with increasing brightness without the selective wavelength filter. When we put the selective wavelength filters on the patients they tolerated the brightness acuity test much more comfortably.

What filters have we found to be the most beneficial? We are working on correlating and crunching all of the data. I took a quick look. What I found is that so far the majority in the group of mild brain injured soldiers are selecting a 527 nm filter. Next is the 550 nm filter followed by the 450 nm and the 511 nm.

Any of the selective wavelength filter colors used for the patients also got a polarized filter over it for outdoor use. We are getting very creative with these filters, even putting the filter tints into contact lenses. The soldiers get very self-conscious about appearance with their red-orange-yellow filter glasses. This is one of our soldiers without lenses (picture shown during meeting). You can see his natural eye color. He has contact lenses with a 550 nm spectrum filter tint with clear corrective glasses over. The contact lenses we use are generally made without prescription, which allows the lenses to last longer since the contacts are custom-made and we put prescription glasses on top for the refractive correction; however, prescription contacts can also be custom tinted.

So that is the clinician's perspective on what we are doing with binocular vision, accommodation, and photophobia problems in mTBI patients in a quick nutshell.

Discussion

>> AUDIENCE MEMBER:

We found the DEM correlated with oculomotility and used it on all the patients. It is a good, inexpensive tool, and also a clinical tool if used properly. A lot of people do not use the tool too well, but it is a good tool to make the diagnosis and thereafter look at treatment, whether it is vision therapy or prisms.

>> SUZANNE WICKUM:

We are looking at the King-Devick as a screening tool. It is not just looking at eye movements, but also at cognitive issues because obviously someone has to see a number, recognize the number, and then say the number. So there are multiple components to the test, but those components all tend to be affected in brain injury. It is a promising screening tool for knowing when to sideline someone with a suspected concussion.

>> AUDIENCE MEMBER:

With such a large battery of tests you are looking at so many parameters. Do you worry a little bit that at some point you will get a high percentage who will be abnormal in one or two tests by just chance alone? If people looked at enough normal controls compared to the TBI population and looked at the correlation of the abnormal test results with the symptomatology this would give us a sense of confidence that we are not picking up stuff by accident.

>> SUZANNE WICKUM:

I do not do every one of these tests on every single one of my patients. I am picking and choosing from one case to the next. Is the phoria status abnormal? If so, is there adequate vergence ability to compensate? I am looking for trends in the data. Are there multiple data pieces that all concur that one entity is present? Or is the data all over the place and inconclusive?

Also, it took me a while to realize that many soldiers are not always super-honest about medications. We would start some therapies with them and they would swear they were doing the therapy. We had physical therapists and occupational therapists reminding patients to do their home therapy, but they would come back no better than before. Some of the data would be all over the place. There was nothing consistent for a diagnosis.

As we probed more, two things came across. The soldiers would say, "Oh, I did not tell you, Doc, I was still hung over the day you worked me up." The other thing is that a number of them are using marijuana on the side; it is something they are doing on their own. They say it is alleviating a lot of the other psychiatric symptoms that they have and has fewer side effects than some of the doctor prescribed pharmaceuticals.

Once we get the truth out of them, we can approach treatment from another standpoint.

>> AUDIENCE MEMBER:

Do you see much convergence spasm in your population? We see it sometimes and we do not know what to do so I do not treat it. I come across it every now and then.

>> SUZANNE WICKUM:

A couple of the PT's were just asking me about that two weeks ago. I really do not come across it very often. Do the other clinicians in the room see it?

>> ANNE MUCHA:

My population is sports concussion and we see a lots and lots. We usually identify it with infrared goggles. Sometimes it is positionally induced. It is pretty interesting. Usually there are some symptomatologies that correspond to it. Probably 1 in 10 concussion patients will show disorders.

>> AUDIENCE MEMBER:

How soon after the injury are you seeing them?

>> ANNE MUCHA:

Typically after the first two to three weeks. In theory, it is a disorder of divergence. It may also work for spasms as well the control issues. If it is severe, we will send them for cycloplegic intervention.

In theory and in various cases we have explored Botox. By and large, these are not to the level where we really need to intervene with medication. We do not know. That is a part of the reason why I am here.

>> SUZANNE WICKUM:

Do you know if they were having accommodative spasm too, because we know that the young population often suffer from accommodative spasm after TBI. The accommodative spasm could contribute to convergence spasm through the near triad.

>> ANNE MUCHA:

I do not have any numbers on that per se. I will say that we see the patients that might have convergence spasm. A high number has vergence inconsistency coming with it as well. It is interesting to try to discern what to do.

>> AUDIENCE MEMBER:

When you get all of this data on all these people do you use it to try to determine what kind of eye damage they have sustained, especially regarding eye movements?

>> SUZANNE WICKUM:

I have not yet. I'm a clinician in the trenches. Recently some of my colleagues at the university are trying to collaborate with us, following through on ideas stemming from our clinical experience. What we are seeing and getting them to help us look at correlates to specific pathways. There are other groups that are trying to tease out different pathways.

>> WILLIAM GOOD:

Our next speaker, Christopher Tyler, needs no introduction to those of you who work at Smith-Kettlewell. He comes from England. He began his research career at Keele University in the North of England and has been in the U.S. I think 40 years. Most of that time has been with Smith-Kettlewell. He is a research-on-demand person. We are very interested in Functional Imaging for Oculomotor Deficits in mTBI.

Functional Imaging for Oculomotor Deficits in mTBI

CHRISTOPHER TYLER, PhD, DSc

I hope it is not a surprise to everyone that I come from England. I do not think I have lost much of my English accent. I am very happy that we did get funding to study oculomotor problems in mTBI. We are really very appreciative of this funding, and you'll note all of my team are here in the room.

Oculomotor Sequelae of mTBI

What happens when someone hits you on the head? Things get very fuzzy and the eyes tend to converge. Convergence is a very strong symptom of the traumatic event. Although from interrogating people in the field, I have not found anybody who has analyzed that symptom specifically as to the pathway. This was the inspiration for our study and the talk I gave earlier was an attempt to investigate the mechanisms from a global perspective in the human brain.

This is the correlation that I talked about between the pattern of stress in football concussions and the long-term tissue shrinkage. You can see that these patterns match very well, suggesting that this is really a clue to the mechanism that is involved. The big factor here that I'm going to focus on is the predominance of the tissue damage in the brainstem, which is the site of control of ocular convergence. This gives a clue to the pathway that may be involved, why convergence might be a symptom of TBI.

What we did to go after this question was to look at the brainstem in detail from the viewpoint of functional MRI. The brainstem and midbrain contain the cranial nerves. They are the oculomotor, trochlear, trigeminal, abducens, and facial nerves, of which 3 are ocular control nuclei. I've put together a diagram of the convergence pathway because I could not find anyone who summarized it to quite this extent.

The degree to which these pontine and midbrain nuclei are involved in oculomotor control is shown by this circuit, which is largely derived from the work of Paul Gamlin (see Gamlin, 2002). There is a cortical signal to converge the eyes. The signal travels down to the pons and then out towards the cerebellum, and then, from there, back out to the supra-oculomotor nucleus in the midbrain. Then it heads to the medial rectus muscles, the convergence muscles that are mediated through this

87

cerebellar-midbrain pathway. The divergence signal travels down through the pons, through the abducens nuclei, to the lateral rectus.

Most of you ophthalmologists are very familiar with this arrangement. The idea that the two different muscles on the two sides of the eye are controlled at entirely different levels in the brainstem strikes me as very surprising, from an evolutionary perspective. But that appears to be its organization.

Functional Imaging of the Brainstem Oculomotor Nuclei

What we wanted to do was to go after this organization from the point of view of functional activation using Magnetic Resonance Imaging (fMRI). To achieve this, we developed a unique high-resolution prescription that targets the brainstem region. The anatomical T2 image shows the resolution we could record for those who are interested in this brainstem/midbrain region. I will be talking about three main nuclei: the superior colliculus, the supra-oculomotor and the abducens.

What kind of activations can we get in these regions? The superior colliculus is the control region for eye movements. It sits at the top of the midbrain. We found activation corresponding to a vergence paradigm. The subject in the scanner had to make convergence and divergence eye movements for a 20-second period and then rest. They basically repeat that cycle for about half an hour.

The kinds of signals we get from this paradigm have the expected time course of fMRI signals. When the vergence period begins, the signal in the superior colliculus region increases. Then it maintains a steady level. After the vergence stops, in the rest period, the signal declines again. This is how we can record an fMRI signal in this superior colliculus region.

I need to point out that, while the subject is making the movements, they are tracking a target, which is thus jostling about on the fovea during the vergence period, but there was no significant visual stimulation during the rest period. Thus is not unexpected that we get strong activation in the lateral geniculate nucleus, which is in the visual pathway from the retina to the cortex.

Less commonly recorded is activation to eye movement signals in the cerebellum. The vermis is the central worm-like structure along the middle of the cerebellum. We find a strong fMRI signal in the vermis, which is a non-visual area. It is an oculomotor control area so it is not surprising to find a strong fMRI signal for the eye movements. On the other hand, a very challenging nucleus to get at is the abducens. It is at the caudal end of the pons. We have to identify it by location cues because there might be a couple of noise spots and there also may be other visual activation.

But when you see it bilaterally, it is very convincing that this is the abducens nucleus. We do find it has a rather noisy signal, because it is such a small nucleus, but we did feel we were able to get fMRI signals from the abducens.

Finally, the supra-oculomotor nucleus is just below the superior colliculus. Again, we were able to get a good fMRI signal in this nucleus. This is a proof of principle that, for the first time, we are able to really record both signals at this high resolution from these brainstem nuclei involved in the vergence control pathways. This is an overview of the results we are able to get.

I would like to integrate these results with what I covered earlier this morning: that a high proportion of mTBI patients do show oculomotor deficits, as confirmed from the work of Greg Goodrich and all the VA folks.

Also, our analysis of the key regions of mTBI damage identify the primary involvement of subcortical and brainstem loci housing the oculomotor nuclei in particular. That derives from the methodology we have developed for functional imaging of this network of oculomotor vergence control nuclei.

However, I should point out this is only one of many potential applications of functional MRI in traumatic brain injury. There are many other directions one could take.

Discussion

>> AUDIENCE MEMBER:

How long does it take to run that scan on the brainstem? Is this your normal control group?

>> CHRISTOPHER TYLER:

This is a pilot study. I showed you our best subject. It was a one-hour scan.

>> AUDIENCE MEMBER:

Is it affected by claustrophobia?

>> CHRISTOPHER TYLER:

Claustrophobia affect is certainly a possibility. In ten years we have only had a couple of people refuse to do it. It is actually a very calm environment once you get into the situation. There are people who like to go into the scanners and meditate.

>> AUDIENCE MEMBER:

A big a part of the problem is being able to fix on the target hoping you will get the output.

>> CHRISTOPHER TYLER:

They have to be able to follow the instructions. We will give them some practice before going into the scanner. You do not want to just throw them into the situation.

Once they are in the scanner they are out of your control so you want to be sure they know what they are doing. You can monitor the eye movements and talk to them, of course, but it is not nearly as tight a coupling as working within the lab with somebody. These are issues that could influence the results.

>> AUDIENCE MEMBER:

Have you had a chance to try this imaging with a few individuals who have sort of a lifelong congenital stationary night-blindness or strabismus, individuals who are not dealing with the multiple injuries of traumatic brain injury? Looking at something a little more pure and reproducible in the long-term basis before going to the more challenging patients with more severe TBI seems worthwhile.

>> CHRISTOPHER TYLER:

 I have worked with strabismus, as lots of people have.

>> AUDIENCE MEMBER:

Are the same nuclei involved?

>> CHRISTOPHER TYLER:

I am showing you the first data, but you would not expect them to be particularly affected in amblyopia. Possibly this would be the case in strabismus.

>> AUDIENCE MEMBER:

It looks like the main issue is noise?

>> CHRISTOPHER TYLER:

What I was showing you is that the signal-to-noise ratios are stunningly good.

>> AUDIENCE MEMBER:

You mentioned that bilateral activation was a huge problem. Isn't that noise?

>> CHRISTOPHER TYLER:

That was in the abducens. The very smallest oculomotor nucleus.

>> AUDIENCE MEMBER:

As far as I understand it should only respond to convergence, not divergence?

>> CHRISTOPHER TYLER:

No, divergence.

>> AUDIENCE MEMBER:

It looked like a better control than just nothing.

>> CHRISTOPHER TYLER:

Yes. What I showed you was a block design. You could do an event-related test also. You would first separate each convergence or divergence in time. Then you accumulate them separately and look for the differential signals.

>> WILLIAM GOOD:

I see certain older patients with divergence insufficiency – How would it be if I brought you -- one of those patients?

>> CHRISTOPHER TYLER:

We are in the early days at this point. This study is to validate the capability of doing the fMRI. I do not want to get into running any patients until we have the basic data.

>> AUDIENCE MEMBER:

If I understand correctly you do not have a lot of data yet.

>> CHRISTOPHER TYLER:

These are the first normal data - proof of principle that you can do it at all. So we are starting that whole concept. I should mention that the DoD grant is also to study the norms of oculomotor control. We have eight different vergence and reading tests. We're going to run 100 normals to find out the normative behavior. As Greg is always saying, we do not really know what the norms are. We need to see what kinds of errors in binocular control normal people show before we can really tell whether a TBI patient is abnormal.

>> AUDIENCE MEMBER:

What do you expect to find?

>> CHRISTOPHER TYLER:

Well, it is a part of the vergence control pathway. We are expecting to find deficits in that signal. When we do the event-related analysis, we can start to look at the sequence of activation through the pathway.

>> AUDIENCE MEMBER:

I was going to follow up to Bill's question. Convergence insufficiency is a very common problem even outside of the TBI. It would be very interesting to see if you can anatomically tease out what the anatomical basis of convergence insufficiency is in otherwise normal and convergence patients.

>> CHRISTOPHER TYLER:

Right. It would offer some striking comparisons.

>> AUDIENCE MEMBER:

I'm sure Pia Hoenig's group could provide some of those patients.

>> CHRISTOPHER TYLER:

We're hoping that might be the case.

>> WILLIAM GOOD:

Our last speaker this afternoon is Anne Mucha. She works with Dr. Mark Lovell of the University of Pennsylvania in the diagnosis mTBI and its rehabilitation post-injury. I'm really intrigued by your title, Anne. I am hoping that your talk on "Concussion diagnosis" can finally bring to rest what is an unsolved problem.

Concussion Diagnosis

ANNE MUCHA, PT, DPT, MS, NCS

This has been a great conference already. As a physical therapist, I am certainly no expert in eye movements and ocular function. However, I am able to speak to the diagnosis of concussion and some concepts regarding management that we utilize in Pittsburgh in our sports concussion program. Sports concussions, like military concussions and blast injuries are another large area of study. Sports injuries number up to 3.8 million per year according to the latest figures available from the CDC. We are talking about huge numbers here. We have a lot to think about.

How we now define concussion is a great change from earlier definitions. This definition takes into account the complexities that we see and that we have talked about all day. Concussion is a complex pathophysiological process that leads to a constellation of symptoms, which are physical, somatic, cognitive, emotional, and sleep-related in nature. Rapid recovery is not guaranteed in this population. It may take minutes, days, weeks, and some patients never fully recover, as we all know. The general caveats about management of mild TBI are apparent here too. The traditional grading systems that have been used - grade one, two, three - or even simple versus complex - have been shown to be ineffective in diagnosing and managing concussion. Unfortunately, we are not in a place where imaging has a role in identifying when a concussion has been sustained and how to manage it. Self-reports, which are a good concept in theory, as I think we've alluded to today, are not reliable enough to direct management or to detect a concussion. Not only is there tremendous variability among those injured, there is also great variability in clinical management and recommendations following a concussion. There is also a lack of education in the general population about concussion, even though every time you turn on the television or open the newspaper, it seems that concussion is a highlighted story. The public perception of concussion is fraught with misinformation.

One of the realities that we deal with in the real world out there is that emergency rooms and trauma departments often do not diagnose and manage direct concussion well. In Pittsburgh we are fortunate to have a wonderful state-of-the-art facility. The Pittsburgh Steelers practice field is on our campus, and part of our building kind of sits beyond it. We also have the practice facilities for the Pittsburgh Panthers on site. And the Pittsburgh Steelers are here in the building in the forefront. We have access to a large population of concussed athletes, with over 10,000 patient visits per year.

As with many things in western Pennsylvania, our program started with football. Our model for objectively diagnosing and managing concussion developed in the mid 80s after former Pittsburgh Steelers' head coach, Chuck Noll. He challenged his medical staff to develop objective measures to determine when his players had recovered from concussions. At that time, practices regarding return to play were based on traditional concussion grading systems and were not evidence-based. His question led team physicians to pursue development of neurocognitive performance measurement for professional athletes. This, in turn, led to the development of the ImPACT computerized testing program, which is now used throughout the NFL as well as in most other major professional sports leagues, collegiate sports and amateur youth sports internationally.

Testing for Concussion

I am going to talk about the development of the ImPACT test, the computerized neurocognitive testing program that we use in our concussion program. We realize that self-reports are subject to quite a bit of error, the largest issues being underreporting, at least in the athletic population. Fifty percent of high school athletes fail to report their injuries. Hockey is probably the worst. There are very legitimate reasons for this. Players and athletes do not want to leave the field. In general, they may not be aware of their injuries. Whenever you're talking about more traumatic events with more severe injuries, there is quite a bit of room for errors. We need objective measures.

In the past 10 years, there has been an explosion of literature devoted to computerized platforms for neurocognitive testing. There are many platforms out there - CogSport, Headminders, ANAM, ImPACT - but the entire concept of neurocognitive testing promotes an approach to detecting concussion and monitoring progress and recovery that is individualized.

The 20-minute ImPACT test designed to look at multiple aspects of cognitive functioning. During this brief test, there are six subtests including verbal and visual memory, visual motor processing speed, reaction time, and cognitive efficiency. There is also a self-report section called the Post-

Concussion System Scale that adds subjective information to the neurocognitive testing. One should not be used in place of the other.

One of the things we see clinically, and the literature strongly supports, is that the younger you are when you're concussed, the longer the recovery will take.

Some question if there is a need neurocognitive testing; and whether it is possible to rely on resolution of symptoms to determine recovery. A recent study by Sandel et al. (2011) looked at the ability of young concussed athletes to estimate their level of recovery. Patient perception was correlated with their performance on the ImPACT test. The correlation back to their actual performance was pretty poor, with symptom report having the worst association with objective information. In another study, this one by Fazio et al. (2007), the utility of neurocognitive testing for asymptomatic concussed athletes was researched. The athletes were divided into three groups and tested within 4 days of injury. The blue bar represents athletes who had sustained a concussion and were still symptomatic. The red group is comprised of athletes who sustained a concussion but were asymptomatic. There was also a control group, which, of course, is performing normally. As expected, when the ImPACT test was administered, deficits in both verbal memory and visual memory were found in the symptomatic concussed group. However, deficits were also found in the concussed asymptomatic group as well. It appears that even in athletes reporting no symptoms, there may still be neurocognitive findings indicating incomplete recovery from injury.

Our program model is one in which we use the baseline system testing when at all possible. Typically, the participants involved in our program have taken a baseline ImPACT test, which is performed in the pre-season. If athletes sustain a concussion, they are removed from play and there is a follow-up process during which they are re-evaluated. This is then followed up by the clinic weekly or biweekly, as needed, until they return back to full functioning or are cleared to return to play.

Several studies have shown ImPACT to have good stability, good test-retest reliability (Schatz, 2010; Elbin, Schatz & Covassin, 2011) and to be both sensitive and specific in diagnosing concussion in sports (Broglio et al., 2007; Schatz et al., 2006).

Conclusion

What I want to impart to you is that neurocognitive testing should have a larger role in diagnosing and managing concussion. We use this test and we recommend some form of neurocognitive testing if you are in the business of working with concussed patients. However, neurocognitive testing is not a stand-alone measure of the injury. It is a tool, along with many others, that we use in diagnosing and managing a concussion.

ImPACT also helps to us manage activity level. Many of the concussed patients we see are student athletes who need to have recommendations about whether they should stay in school or need academic accommodations. They need to know whether they should be allowed to have any level of physical exertion or return to play. ImPACT is also a tool to demonstrate to athletes, families and coaches the degree of cognitive impairment from injury because there are numerical values that can be compared to baseline function or norms.

Probably one the biggest misperceptions is that ImPACT testing is done in isolation. In our program, the neuropsychologist acts as the point of entry, because of the neurocognitive testing piece of it. From there, a screening process is conducted, where then other members of the team are identified and brought into the picture. These could include medical therapy, vestibular therapy, physical therapy (for sports-specific training), neuro-ophthalmology, neuro-optometry, primary care, orthopedics, neurology, neuroradiology, and other specialties.

There are a lot of pieces to our puzzle. Just briefly, our clinical evaluation includes ImPACT for neurocognitive assessment. There is also a detailed clinical exam including a structured interview, an oculomotor screen, and a vestibular screen. I think I will stop there just to allow us to move forward and take questions.

Discussion

>> AUDIENCE MEMBER:

What is the test for those who received what presumably is the first or second concussion? How long do they take to recover?

>> *ANNE MUCHA:*

If I understand your question correctly, typically 80% of the athletes, especially those in high school, will recover within the first three weeks after a concussion. The other 20% take quite a bit longer. They do not recover spontaneously in a period of three weeks, as is shown pretty consistently in the literature. They are the ones for whom we need to go back and involve all the other disciplines.

>> *AUDIENCE MEMBER:*

What happens to the other 20%?

>> *ANNE MUCHA:*

We intervene. They may need therapeutic management or many other services. It is a constellation of things.

>> *AUDIENCE MEMBER:*

Can you take a minute to describe what is involved in conducting the neurocognitive testing for ImPACT? How long does it take? How is it? How many questions involved?

>> *ANNE MUCHA:*

The test takes, give or take, about 20 minutes to complete. It is a computerized test with a subset of different cognitive tasks. I am not the developer of the program, it was developed by my colleagues, and directors of our Concussion Program, Mark Lovell, Micky Collins and Joseph Maroon. There are six different modules including tasks of word memory, design memory, X's and O's, symbol matching, color matching and three-letter memory. In addition, there is a post-concussion symptom scale, in which 22 different symptoms are rated on a 0-6 Likert scale. Symptoms such as dizziness, headaches, light sensitivity and the like are rated as part of this scale.

Again, most of the athletes that we see have baseline information. We will have what their normal function is and in all of those domain areas. Based on the six areas that are tested, four composites scores are recorded. One is visual memory; the others are verbal memory; reaction time and visual motor processing speed

>> *AUDIENCE MEMBER:*

Is this proprietary; do you own it? License it?

>> ANNE MUCHA:

ImPACT operates is its own separate corporation. The owners are Mark Lovell, Michael Collins and Joseph Maroon. We use ImPACT, though, in our cognitive evaluation.

>> AUDIENCE MEMBER:

You said very few of the people have loss of function.

>> ANNE MUCHA:

Less than 10 percent.

>> AUDIENCE MEMBER:

What is the criterion that you have a concussion? What is the thing that makes a head impact into a concussion event?

>> AUDIENCE MEMBER:

When they run to the wrong team's side of the field.

>> ANNE MUCHA:

This is actually where the debate is. How do you define a concussion? And that is why there were sideline assessments of concussion and forms used by the athletic trainers.

>> AUDIENCE MEMBER:

What is the general concept? What begins the discussion?

>> AUDIENCE MEMBER:

If they fall down and do not get up.

>> ANNE MUCHA:

Not necessarily. Either the player reports a symptom that they do not typically have, or the athletic trainer will notice a symptom. Sometimes the player appeared dazed or confused. They report a headache immediately. That is not normal. Actually, what is interesting is the report of on-field dizziness. We just put in press a study on this. We found that if dizziness is one to have reported symptoms following an impact, it is the best predictor of a poor or protracted recovery. This is definitely a symptom that is not typical and normal. Most players will know if they feel dizzy, although not all.

>> *AUDIENCE MEMBER:*

Your results sound like the standard rate for people who have concussion but do not want to be diagnosed. What about the opposite, when someone did not have a concussion but claims she does? Will it work for them?

>> *ANNE MUCHA:*

There are things built into this program that look for malingering. Other tests of efforts are usually performed that are paper and pencil based.

>> *AUDIENCE MEMBER:*

Suppose it is psychosomatic. What if someone thinks that he banged his head and therefore he must have a concussion? He might not consciously be trying to cheat, but thinks he has one.

>> *ANNE MUCHA:*

I do not know that there are any great tests for people who have conversion type symptoms. The more objective we can get, the better. This is why we use vestibular and other balance tests. We do oculomotor screening.

>> *WILLIAM GOOD:*

This might be a very good question for the panel to study. How come you have 10,000 concussions per year at Pittsburgh?

>> *ANNE MUCHA:*

There are 10,000 visits a year, which includes new and returning patients. While the majority of our patients are from Western Pennsylvania, we also see many patients and athletes from across the US.

Panel Discussion

Arthur Jampolsky (Moderator), Randy Kardon, Michael Gorin, Gregory Goodrich, Suzanne Wickum, Christopher Tyler, Anne Mucha, Yury Petrov

>> ARTHUR JAMPOLSKY:

First, I want to ask the panelists if they would like to ask questions of each other? They are the experts in their fields. We certainly covered a panorama today, from one aspect of end of spectrum to the other end. It was a fascinating afternoon. Chris you always have a question.

>> CHRISTOPHER TYLER:

I wanted to pick up the question of the precipitating event for TBI. I am not clear about the difference between blast injury and concussion. Is it air or a solid object that did the damage? Randy?

>> RANDY KARDON:

Although, obviously, both the dynamics of the force and how it interacts with tissue are factors, there may be some overlapping similarities. The problem is that animal models of concussive injury are highly variable. This is why scientists working with the impact concussive model are also seeking other models of TBI. Most people do not lump both models of TBI and separate them to study different mechanisms of damage. However, many cases of TBI in humans are a mixed bag with both concussive and blast-wave injury to tissue.

It is correct to say that concussion is a mechanical damage interaction with something hard. The other mechanism (i.e., blast injury) would be more like a pressure change as the blast wave passes through tissue, which may cause different degrees of damage depending on the water and air content of tissue and the characteristics of the pressure wave. This would cause variable damage in tissues such as the lungs, middle ear, eye, blood, and brain.

Some of our colleagues from the DoD may want to correct me, but my understanding of blast events is they can be very complex. You have the blast wave itself, which may leave you flat on the ground. Your head may also hit the ground or a hard object as a result of the blast event.

100

In combat situations, the protective armor and the vest may influence the effect of the blast wave and the other physical trauma that can occur. If you are in a vehicle your head can hit the vehicle. If you are in a building, you may hit the wall etc. We do not want to think of a blast over-simplistically. It is good to look at a small-animal model for concussion, but it is apparent that more complex things happen in the real world and how we attempt to thoughtfully translate from research to people that have been injured is an important aspect of the research.

Part of the results of analysis of human responses to concussion versus mild TBI may depend as much on who is evaluating the patient, taking a history and writing down whether they interpreted the patient's experience as a concussion or mTBI. Some clinicians do not want to put the label of mild Traumatic Brain Injury onto a person. They feel the label attaches an unnecessary stigma to the patient. There are also various rating systems to classify a concussion and its severity, which are very dependent upon the criteria that a clinician applies to the patient's experience and their physical response to the initial injury. So, on one level, I think this whole discussion of concussion versus mTBI is a very important one, but very murky, and . . .

>> *CHRISTOPHER TYLER:*

. . . unresolvable.

>> *RANDY KARDON:*

It may end up being executive fiat.

>> *FRANCIS McVEIGH:*

A general concept is talking about the transmission of energy to the brain. We are talking about brain and about energy going in one way or the other through direct force or a blast wave. The unifying thing is a sudden incremental amount of energy. I said, well, what if I applied electroshock to someone, would it be part of that discussion, too? Energy into the brain. I would not exclude that either.

So in one sense, we can talk about what is the mechanism by which the energy is being delivered. Is it being delivered diffusively or focally? Or, is there a sort of time course? You can parse it out that way or look in the global sense.

There are some advantages to looking at it in the global sense. What sort of brain injury do we cause when giving electroconvulsive shocks? That is something to perhaps think about so as to not make the definition overly limited to the mechanism by which the energy is applied. We could recognize that it can be also be ultrasound. This is probably more universal.

>> AUDIENCE MEMBER:

You would not want to exclude those people from being studied.

>> FRANCIS McVEIGH:

We belabor these definitions all time. One of the things we say in our arena is that we want to know the visual affects associated with blast ballistic. Ballistic is one that we often see and it is not unusual to have a penetrating injury to the brain as a singular event. Then, what happens to the localized concussive events, the energy transmission events? All of these become elements of brain injuries. We try to shy away from a definition of traumatic brain injury, because the classification does not always relate to the event. At least we have not found a design yet that allows us to say: you have this or you have that.

I do think we are approaching some definitions. All of the strategies that we are talking about now involve classifying the degree of damage and the location of damage and the consequences of the degree and location damage. But symptoms do not always relate to the degree of damage. That is what you are finding? You are using symptomatic evidence, which may not necessarily correlate with traumatic brain injury, or with brain damage.

One of the most intriguing parts to the work we are looking at is multiple low-level blast exposures. We do not have a good mechanism yet to understand the event and the correlation of the event with the outcome. We probably do more in the ballistic side. We all know what happened to Gabrielle Giffords. We can clearly document that event, the outcome of that event, and what damage was done. In the blast, especially the low energy open field blast, there may have only been a feeling that you had some energy thing pass through you. That happens five, ten, 15 times and suddenly you lost the ability to have memories. It is clearly related to brain injury and repetitive brain injury in some specific areas that cause diffuse axonal damage.

>> MICHAEL GORIN:

On the furthest end spectrum, you could even consider the cell phone.

The latest report that came out said that there were brain alterations in the temporal areas. This in not necessarily saying there is injury, but there is a very frequent low energy exposure. I think it is wise to consider the entire spectrum of energy delivery to the brain.

>> ARTHUR JAMPOLSKY:

Thanks. I did not realize that your time here at Smith-Kettlewell would be the foundation for such a brilliant career. It has been very, very varied and I am pleased to know you. Any other comments on that definition? I guess we are going to fall flat on a definition?

>> AUDIENCE MEMBER:

If we cannot agree on definitions for mTBI, concussion, and blast injury, what about mTBI and moderate TBI? Are these more or better defined?

>> ANNE MUCHA:

I think we use the Glasgow Coma Scale to look at typical TBI. This provides a pretty objective way of looking at the more severe brain injuries.

>> ARTHUR JAMPOLSKY:

I want to repeat a question that was answered, one that was proposed before. You take a heck of a lot of tests. How do you use them? Do you pick one as the most informative or do you put them all together and divide by X? I can give you an orthoptic history of which I was part. You take a lot of tests and then the surgeon does what he damn well pleases. Really, this is not a joke.

Most of the orthoptic findings on this major amblyoscope were put aside in the final analysis. The surgeon proposes what he wants to do which follow his bias. Is that a fair analogy? I do not think so, but let me use it anyway. Which test do you use? Of all of the tests you administer, which ones are you going to use to evaluate something that will allow you to manage the case?

>> MICHAEL GORIN:

Well, it is a chicken-and-egg type of process. In genetics we do tons or hundreds of thousands of tests on a person and then we decide what is important. At this stage of our knowledge, you probably run a lot of different tests, look at the outcomes, see which tests do the best job, and then do a replication process to see which tests actually hold up. I do not think we are at a stage now where we can point to a lot of these things.

There is no gold standard, Art. I think the only way to do it is to try a lot of different tests, look for cross validation and see which tests correlate with others. The bottom line is, what goes along with the outcome?

>> YURY PETROV:

When you have some data and you want to test the model. In this case, the model would be when the subject's dysfunction is not based on the data. Statistically, there is a mechanism on how this can be done. The only problem is getting enough information from normal subjects. But, what is normal? The problem is building the distribution for normal subject. It is better to use more tests and more different tests for more informative distributions. It is very straightforward statistically to invert it and to decide whether a particular subject or patient has the condition or not. In my view, the only thing that needs to be done is to get more and more different kinds of tests on normals. It is just a matter of time.

>> CHRISTOPHER TYLER:

If you do a lot of tests, you are bound to find something positive eventually. It is an ill-informed procedure. I also heard this idea expressed from the Dean of the Stanford Medical School. I would still characterize it as ill-informed.

When you apply many more tests, you need to raise what is called the protection level of the statistical criterion in proportion to the number of tests. Then you should not get any more false positives from multiple tests than you did from one test. If you apply that criterion, you are still safe; but you can often pick the ones that pass that criterion and give you the significant result.

>> MICHAEL GORIN:

In genetics we have moved away from testing one gene or variant at a time, to investigating many genetic variants in a single experiment. The whole key thing in genetics these days is replication. After you find something of interest, whether it is above some certain threshold that you set in order to pique your interest, you then go on and try to replicate it. The problem with a lot of these studies that we're doing these days is that they are so costly. The idea that you will replicate your results from 500 TBI cases and 500 normals with 1500 TBI cases and 1500 normals is inconceivable from the government funding point of view. We just take off from what we have done and go with it until someone disproves it.

Part of the problem is our experimental designs in certain areas of study. We have not really embraced the fact that there is this expensive secondary process that has to be done after we've done these widespread surveys for different types of either testing parameters or therapies.

>> YURY PETROV:

Returning to Chris's demand, this view was solved in the 17th century by Bayes. It is really not a problem. Again, the larger problem, as was mentioned, is to get enough statistical data to get a good distribution so that we can understand what really normal means with all these tests.

>> ARTHUR JAMPOLSKY:

I am going to turn the discussion over to the audience now, particularly our military and Veterans Administration guests. Are there any questions you would like to pose?

>> ROBERT MAZZOLI:

As someone who is going back to the steering committee to try to expose gaps, do you have any new gaps that will help us drive where we need to go?

>> RANDY KARDON:

What genes are turned off after an injury? Some may be beneficial and some may be harmful. Understanding what kind of genetic profile of transcriptional activity is present in tissue at the time of acute injury compared to later time points might help one gain an understanding of what different treatments may modulate the response to injury to help minimize the net effect. This may provide another avenue of formulating new treatments for TBI.

>> CHRISTOPHER TYLER:

I am going to be predictable and say functional imaging is a major gap. There is almost nothing done. It has to be targeted to the specific conditions that you suspect. It may be a problem to do so, but this is exactly the nature of a gap.

>> GREGORY GOODRICH:

I would like to get estimates of the magnitude. How many soldiers and sailors are affected? If we have a few hundred thousand people that have been exposed to blasts, and therefore, possibly have an mTBI, that really tells us the problem could be huge. But it may not be. I think we need to have a better idea and understanding of the extent of the problem. I also think we need it in the civilian world. These kinds of numbers are potentially huge.

I think understanding the epidemiology of the problem, so to speak is important. The other side that concerns me is binocular dysfunctions. I have no doubt that some people with binocular dysfunctions have difficulty reading, but do all the people that we identify as having a binocular dysfunction have a problem? So, what is the extent of that problem? How big of a problem is it?

Is it something that some people might think of as trivial in their lives, while other people find it a major problem? I recently had a cataract removed in one eye. The irony is that I have spent my career looking at people with real impairments and the last three four months, when I had the cataract, I was bitching and moaning even though it is not a huge issue the way things go. But it showed me that binocular dysfunctions would be a major problem. I do not know if this generalizes to the entire population of people with binocular dysfunction. Those are two kinds of things I would identify as gaps.

>> MICHAEL GORIN:

You could have people who objectively have the same degree of impairment. One is not bothered at all but the other one is literally disabled from his or her career because of the nature of his or her work. This is an interesting issue.

I want to reiterate what Randy said looking at the changes in the modulation of pupillary responses as a function of age. I am impressed that younger people take longer to recover than older people. This means to me that there are changes going on that in our own bodies that alter with age. It also gives me the thought there are areas where we can intervene particularly in the case of the military where they are young people getting hurt. There may in fact be ways of using the knowledge that younger people do take longer to guide our work. We could maybe work with younger and older animal models to find alterations that are more severe in the younger ones. This could help to modify the extent of damage and hasten the healing process. Is this a gap we could address in a straightforward way?

>> KENNETH CIUFFREDA:

I wonder about looking for markers of the occurrence of the concussion, the electrophysiological aspects of this. We did a study in children with diffuse brain injuries and found symptom reduction if you tested under binasal occlusion. We also heard from the basic science people this morning. We have identified there is ganglion cell loss. There are ways to tease out the phenomenon so you have a real physiological marker for somebody who had a real injury.

>> AUDIENCE MEMBER:

Physiological markers are a huge focus, as both the diagnostic and the neuro-protective or neuro-interventive probe.

>> AUDIENCE MEMBER:

You mentioned doing 20 different tests only to find the one or two positive results. Everybody's protocol talks about finding the genetic marker. It always starts off by saying we will run an array of 60 tests, we will run a bunch of bioassays and focus on the 20 best ones. This is why we were doing 20 different tests to find the one positive result. But genetic markers and biomarkers are clearly an important focus.

>> ARTHUR JAMPOLSKY:

Let's assume it is a real problem and since the criterion for getting help is not necessarily that you show that you have had some exposure to a real concussion or a blast. Rather, you merely say so. That is going to be a real gap in some way. The question that was asked here before and discussed a little bit is not malingering. These people could be ill from depression, et cetera, et cetera.

How do you rule out any brain damage, etc, without going through all the tests? That is a future problem and I have talked with people about what to do about it? How do you rule out somebody who claims they really do not need all the tests? You are not going to run them all on everybody? The Government would go broke.

>> AUDIENCE MEMBER:

Probably from theater, sending somebody back to the field of play or field battle has to be sometimes a very quick and accurate decision. Having high sensitivity is critical. I think cognitive tests help, but from a binocular standpoint, I am not sure what those are.

>> *ARTHUR JAMPOLSKY:*

Any others on the panel want to tackle this?

>> *GREGORY GOODRICH:*

It depends on context. You gave a good example. When you are a leader on the battlefield and one of your personnel may have an injury, specificity is not that important. You know what you are looking for, basically can this person perform their job or not, so you are looking for something with the sensitivity to detect it. Your concern is about the immediate mission and that you may be looking at large populations of troops. If you are a clinician in the field you have to worry about both sensitivity and specificity. It depends on where you are heading – are you leading troops on a combat mission or are you in a medical capacity concerned with both fitness for duty and the future health of the patient. You may want to know who has brain injury and whether it is due to TBI or not, but depending upon the context of the field commander or the medical evaluator the context changes. In the heat of battle it may not matter, but in the longer term you still want to know. Also of importance for the military, is that the some injuries will lead to discharges and if the discharge is medically related it will determine who receives disability payments. And, in my experience, this may be an issue, not at the instant of being discharged, but later in the person's life when the impact of the injury becomes more apparent in civilian life. Still, the causality is as important as the severity of the damage, especially when the impact appears after discharge. Symptoms occurring after discharge raise questions that are often difficult to answer. Is the deficit and resulting liability for it related to an event during military service or if the guy got it because of drunken bar fight after he got out of service? Or did the bar fight exacerbate the TBI that was incurred during combat? This, obviously, is not my area of expertise, however I think it is important to raise these concerns.

It is all very context-driven. You really cannot look at this thing solely in one way. It all depends on the circumstances. In developing comprehensive strategies you have to do an ROC-type analysis and ask yourself under what context you will be using it.

>> *AUDIENCE MEMBER:*

A clinical test has exist to rule out what you want to call visual field defects and will produce false negatives or defects with a total blackout of one field. In my experience, it is one field that is blacked out and can still function. We occasionally see this, and I have seen it in the mTBI or TBI population.

We can trick them easily in that context using tools we have had for many years. By that time they have spent time in the system and have perhaps been an inpatient. Do you agree, Randy, that we can catch those?

>> RANDY KARDON:

As you implied, it becomes a big social problem. Showing people have vision problems when they say they do not is just a tip of the iceberg. What you do about it and how you restore a person back to a productive life? That really takes a lot of effort.

One of the other areas I wanted to mention and get other people's opinions on is what is termed rehabilitation? There are a number of philosophies and people who have an emotional allegiance to certain philosophies about what visual rehabilitation is. Important questions are: Is a certain rehabilitation approach effective? What is the evidence that it is effective?

I think that we really need some very hard research both in humans and in animal models to try and understand if there is a way of strengthening preferred neuronal circuits that can be permanent. Is there a treatment that will strengthen the synapses, to allow better function; and is long-lasting? When evaluating specific rehabilitative treatments, it will also be important to determine if they are specific for that modality or whether the treatment crosses into other modalities. Healthcare professionals disagree as to the definition of visual rehabilitation and how effective it is. In this regard, well-designed scientific research is needed. Politically it is a question of what Congress wants and the response from the public sector. Most everyone wants something done, but it is important to only institute rehabilitative strategies that have been found to work in a well-designed placebo trial with specific relevant outcome measures.

We really need some hard scientific evidence about the efficacy of any rehabilitative strategy before spending millions of dollars to implement it. I would like to add that this is an area where the DoD has a real opportunity, in the sense of funding research that will help determine the criteria for what would constitute a desirable rehabilitation outcome. Such research will inevitably filter into the civilian sector as well. If we can establish an accepted methodology for rehabilitative effects and treatment efficacy for any of these behavioral, cognitive, or visual therapies, it would make a huge difference on a national basis for accepted treatments.

>> ARTHUR JAMPOLSKY:

I'd like to ask the people who have a test for binocular vision about the monocular thing he mentioned. If you have abnormal binocular vision and you have symptoms in reading, which is the number one problem? Have you tested reading for the patients with one eye? Is this a reasonable comparison to make if you have no binocular system that is working? What happens if you do that, if that has been done?

>> SUZANNE WICKUM:

I do not know of any formal studies in that area. Clinically, at times, the data may not be conclusive if the patient has a CI or an accommodative issue. We might have them work under monocular conditions for a time and get some subjective input about symptoms. Is it easier or not? Actually, we can get some of that information from some of the therapy teams before we see the patient. One scenario is: you seem to be struggling with both eyes. I went ahead and put a patch on and reading was better. Computer work was better. In this kind of case, when the rehab team sends the patient to us, we know it was likely a binocular deficit contributing to the patient's symptoms. I do not know of any particular studies that have been done on this.

>> ARTHUR JAMPOLSKY:

Any comment?

>> GREGORY GOODRICH:

The bias would be that binocular reading is better than the monocular reading. If you can restore binocularity, there is probably an advantage to that. In some cases, our clinicians working with patients in the Polytrauma Rehabilitation Center work with a severely injured individual in order to facilitate PT and OT. Quite often they will patch. They will occasionally patch an individual because it removes the disparity in sensory information that can lead a person to be unstable due motion sickness and other symptoms. It is not necessarily for reading. It is more to promote physical rehabilitation, the process. I do not know if our clinicians do monocular patching for reading with any of our BV patients as a treatment, it seems more likely that it is done diagnostically or for a short-term alleviation of the problem.

>> KENNETH CIUFFREDA:

We have not done it monocularly to this point. It is difficult to segregate it out. Even if we do cover an eye, so many of them have a concomitant dysfunction versus a vergence dysfunction. It is an interesting observation, however.

I tell my students, half in jest, if your patient has a binocular vision problem and you have to study for a midterm that night, go buy an eye patch for two dollars. It works really well. I got the same response from the students. That is a really good point though. We do it some of the time with different patients. There are so many concomitant ones, the vergent ones. Retrospective studies show that 50% have some sort of AI, CI, or saccadic abnormality, and the clinic visits show vergence, strabismus or CN palsy. It gets a little complicated. Your point is well taken.

>> MICHAEL GORIN:

The question "what's normal?" has come up several times. We do not need to rely on the overwhelming evidence of 3 million subjects that have been tested with a particular test. Rather we can take on a completely different attitude, namely supply data that allows us to make a comparison between the before and after. For instance, in a sports team or a military, it would be entirely possible for researchers of the field to agree on a set of tests. It would have to be something reasonable. I am thinking not 350, but maybe 25 tests that are done on everybody before they are sent onto the field or into combat. In this case, questions like whether this is a pathologic effect that we are looking at after an event or is this something we need to worry about or not, would be much, much easier than they are right now. What does the panel think about this kind of approach?

>> ANNE MUCHA:

So the idea of baselining is, we think, where it is for the high-risk populations with concussion. Right now, the only widespread baselining is in cognitive functioning. I am also a vestibular therapist. So we are talking about performing baseline tests of vestibular function in some athletes, because the degree to which we see these problems post injury. It is difficult, even with normative information, to know how these tests compare to baseline. Obtaining baseline balance measures is becoming more common, and perhaps a few select places measure baseline vestibular function. I do not remember exactly, and this would be more in the concept of the Visagraph eye-tracker, but pursuits and saccadic movements would be good things to baseline. I know that they do baselining of that at the University of Texas.

>> ROBERT MAZZOLI:

To add one question to what Michael said about the concept of post-testing, have you run the ImPACT on everybody at the end of a regular football game?

>> ANNE MUCHA:

We did not conduct this study ourselves, but interestingly, there was a study by Miller et al. (2007) that looked at a collegiate program, as I recall. They looked at non-concussed athletes and looked at the beginning, middle, and post season to see if they stayed stable. Only the concussed athletes showed changes. Is this what you are asking?

>> AUDIENCE MEMBER:

Sometimes it is not practical. We have a volume of patients who are deploying and it would add to all of the things that we put on them now. That would be the gold standard. The second thing I am happy to see. You have helped prioritize the gaps more.

>> AUDIENCE MEMBER:

We cannot figure what to baseline. It would certainly be a great recommendation because we have been trying to figure out what are those easily administered tests that can be easily retested. I know that you are looking at something post-injury that can be easily retested, maybe bilaterally in these patients that have cognitive dysfunction. The other component of this particular population is that there are not isolated dysfunctions. We heard a little bit about auditory-vestibular concomitants and cognitive in the same patients. We want to figure out how to baseline the visual dysfunctions in this grouping of pairs. This is one I'll throw out as a question.

>> AUDIENCE MEMBER:

You are going to have to worry about drug abuse even for your baseline. If you have people who are, let's say, sampling agents that you may not be aware of, it could obviously throw things off. It may also affect their willingness to comply if it became generally known that you could detect subtle cognitive dysfunction related to things that they are not supposed to be doing. This possibility is something that has to be factored into the equation.

Saturday March 5

Session 3: Therapy

Current practices
Glenn Cockerham, MD

Current Practices
Gregory Goodrich, PhD

Accommodative and convergence training in mTBI
Kenneth Ciuffreda, OD, PhD

Medical or ophthalmic interventions in the globe
Kim Cockerham, MD

Comparison with management of hearing impacts
Gabrielle Saunders, PhD

>> *JOHN BRABYN:*

Welcome to the Saturday morning session, focusing on treatment and therapy. Now that we know what the problem is and have explored what tests we want to do to figure it out, what can we actually do to help the people who have the problem? Interestingly, it was hardest to find speakers for this session. Maybe that means this is one of the areas where there is a lot more to do. In any case, we are really lucky to have, as our first speaker, Glenn Cockerham from the VA Palo Alto. He is going to trade off with Greg Goodrich, also from the VA, for the first talk on current management practices.

I would like to add that without Glenn this meeting would not have happened. He was instrumental in helping us identify the speakers, pointing us in the right directions in the literature, and serving as an impetus as we organized this meeting. I would like to propose a special vote of thanks and congratulations to Glenn.

Current Practices

GLENN COCKERHAM, MD

I thought yesterday was a great session. We covered some ground and we talked about things we do not know and a little bit about what we do know. One of the things Dr. Goodrich and I talked about yesterday was the fact that we are finding some dysfunctions and some injuries in this group in the Palo Alto Polytrauma Center. Almost half of those we looked had some objective eye injury, closed globe injury that would not normally have been picked up unless you looked for it.

Dr. Goodrich has talked and published extensively about binocular disorder, convergence and so forth. We are going to do a tag team talk here today.

TBI Diagnosis

In 2008 the VHA came out with Directive 2008-065, which mandated that every prior (since February 2005), current, and future patient with a diagnosis of TBI admitted to a polytrauma center must have a TBI- specific ocular health and visual functioning examination performed by an optometrist or ophthalmologist. I want to go over that exam.

We were pleased that the VA took this stance. I think the Department of Defense is looking at something similar. I cannot say that there is a perfect template, but I think what the VA did was realize that there is a plethora of possible injuries and dysfunctions in this group. They gave us a guideline to go looking for it. The required history of every patient is supposed to include examination by an eye-care provider and a record that includes the history of any trauma.

When we ask them to describe their history, I pay particular attention to whether they were dismounted versus in a vehicle, what type of protective gear, and what type of eye armor they remember. In this way you can go far beyond the basic questions about symptoms of dry eye, decreased night vision, pain, visual field, etc. The examination itself is pretty thorough. I am just going to make a few comments as we go along.

What the Directive does is give you a guideline of what should be done. It does not tell you the test to use. We do not standardize, so it is hard to do research from all four different centers since each is using different methods and different levels of expertise. As a general starting point, it is great that these troops are getting this kind of examination.

Eye Problems

We start with the visual acuity since it is a gold standard for visual function across the world. We do some types of screens. We take a penlight, a muscle light, a tube with a red top. We test confrontation fields. We do what we can at the bedside and try to get them into the eye clinic when they are able to come. All of our exams are wheelchair accessible. A lot of these troops have orthopedic injuries and they are in some kind of a limb fixation device, often with a leg up in the air. It is not so easy to examine them, but we do the best that we can.

Oculomotor function is certainly important. We are doing video nystagmography and we can graph it out. The directive specifically mentions an assessment of cranial nerves III, IV, and VI. In actuality, we're testing nerves I--X. Someone mentioned that they lost their sense of smell. It is hard to test, but we do try to get into that.

We are checking cranial nerve VII. A number of our troops have had unilateral facial injuries through the carotid and it severs the facial nerve. It will usually come back with time. We check nerve VIII through the vestibular ocular reflex, and then sometimes we do IX and X.

For strabismus, we do all of the oculomotor tests. An anterior segment examination is in the required testing. Zone one, again, is external.

We do a very thorough eye examination. The eye trauma publications from Birmingham, Alabama, which is where the eye trauma registry is for the United States, are the gold standard for how to describe eye injuries. The most common finding in our eye injuries is angle recession. It requires gonioscopy. If you only have one eye, you do not have anything to compare it to, but you can get a sense if there is marked recession. In addition, we check intraocular pressure.

The dilated retinal examination is important. In addition to looking at the optic nerve, macula and vitreous, with the stereo lens is, you really need a good examiner who can do a depressed sclera examination because a lot of injuries are going to occur in the retina, including peripheral retina. We have one of the best retinal examiners in Tom Rice. He was at Wilmer and now donates his time to the VA. He is a coauthor of the textbook on retinal detachment. Tom can only do an adequate depressed scleral examination in about 80% with these troops, because many are photophobic and hyperalgesic. They are hypersensitive to touch whether it is ocular or cutaneous. As experienced and patient as Tom is, he just cannot get a good exam of those who are overly sensitive. We feel that everybody should have a retinal examination including the depressed scleral part of it. So that is the baseline exam we are doing on the inpatients.

Therapeutic Strategy

The topic of this talk is current therapy. We treat whatever we find. This is pretty basic. It is treated with whatever level of care we are able to provide, whether at bases at Kandahar or Bagram in Afghanistan, Baltimore or Walter Reed Medical Center, or within the VA. I am talking about things that we treated at Palo Alto. Of courses our colleagues in the military have done this and much more at their levels.

The follow-up visits are important. We see them every three months for the first year or two and then every six months probably indefinitely because the risk of traumatic glaucoma is there. If we cannot get a depressed scleral examination on the initial visit, we try to do that at some point.

Discussion

>> AUDIENCE MEMBER:

Have found anything, in addition to direct gonioscopy images, that is of value to you as a surgeon for gonioscopy?

>> GLENN COCKERHAM:

We have anterior segment imaging capabilities, including OCT Visante, the Pentacam, and ultrabiomicroscopy. I do not think that there is any substitute for a high quality globe examination with a glaucoma person over your shoulder if you have a question. In addition to comparison views, gonioscopy will detect evidence of prior anterior segment hemorrhage, which has a strong association with angle recession. We have tried to document everything we find with digital photography.

>> AUDIENCE MEMBER:

Just to harp on a point mentioned yesterday, when you say baselines, is that after injury?

>> *GLENN COCKERHAM:*

It is our baseline, the VA baseline. They have had other exams before hopefully.

>> *AUDIENCE MEMBER:*

Could you say more about the development of glaucoma after the baseline? What are the things that you would relate to the injury; or is that something that you might assume might have happened anyway?

>> *GLENN COCKERHAM:*

It is related to the injury. The two that we have treated in this group both had hyphema. It was still a closed eye injury. There was no penetrating injury to the eye. Examination of the angle was not possible while a hyphema was still present. One of our patients came in with pressures in the fifties, the other developed pressures into the fifties several months after the injury. Gonioscopy showed extensive angle recession. We have also had traumatic glaucoma, but the vitreous hemorrhage was part of the problem. It is hard to clear red blood cells if you have an angle recession scarring.

>> *AUDIENCE MEMBER:*

There were two glaucoma that you have treated or . . .

>> *GLENN COCKERHAM:*

There were two that we treated out of the 46 that I mentioned.

>> *AUDIENCE MEMBER:*

Could you have a guess at the top three treatments that you have done to these patients, in general?

>> *GLENN COCKERHAM:*

We were seeing them three to four months after their injury. Acute things were treated acutely. These are the intermediate problems that we are dealing with at the VA. Again, one is a hyphema that was sent to us with that diagnosis. A lot of people do not have a history of hyphema. It is very easy to miss as they progress from the casualty care on the battlefield to tertiary care. It is not something that someone is likely to pick up, in the context of multiple major traumas. The trauma team runs trauma management. Oftentimes they do not consult ophthalmology unless there are globe injuries or orbital fractures. I do not disagree with that. Some probably had a hyphema was not picked up initially.

>> *AUDIENCE MEMBER:*

Do you routinely obtain specular microscopy as well? I suspect very few of them have corneal decompensation. You already mentioned that there is a significant loss in the endothelial cells. Have you noticed that?

>> *GLENN COCKERHAM:*

I have not analyzed this thoroughly. One of the things that I was intrigued by is that some cell counts actually increase. Some subjects are 19 - 20 years old. It is different from our cataract population. I do not know what is going to happen long term. I have seen a slight increase. It could be sampling error, but generally our paired cell counts are pretty tight. We do multiple tests at our exams and they are very tight in their numbers. Then, six months to a year later, we get another test. Some have significant increase in their central endothelial cell counts. There are papers that show mitosis in human endothelium. It is one of the things that intrigues me.

>> *JOHN BRABYN:*

Thank you very much. We'll now hand it over to Greg.

Current Practices

GREGORY GOODRICH, Ph.D.

I would like to talk a little bit about the rehabilitation side. We have started using the phrase neurological vision rehabilitation for lack of a better term. Whether this is an adequate term or not, I am not sure.

I have a few disclosures. First, I have no financial interest in anything. I am speaking for myself only and not my employer. These are the people that actually do the work.

My three other disclosures: I wanted to start out talking about the patient referral source. If you build a better mousetrap, people will come. I am not sure we built a better mousetrap, but we did build a mousetrap and people are coming. The referral source is our medical centers, the Polytrauma Rehabilitation Center, the polytrauma network site, and optometry and ophthalmology. We also have other services in the VA, which are in B 102 of 150 VA medical centers. Walter Reed and Bethesda are additional referral sources for us.

When you talk to the patients as they arrive there are some common things that they will ask: How come it took so long for somebody to refer us? Why don't you advertise? And how come I cannot get these services in my community?

Assessment of mTBI Deficits

We start with an optometric assessment. I will go through this briefly because I do not think it will surprise to anyone here. Optical recommendations to rehabilitation can include spectacle corrections, tints, near and distance corrections for many of the problems that we have discussed. They can also recommend vision rehabilitation as far as training with adaptive devices, oculomotor therapy, glare vision pattern, orientation and mobility, compensatory scanning, and a couple of things that we did not discuss. Many of these relate to visual perception, visual glare deficits that occur in this population.

Vision perception is a very big issue for many of these patients. It can be visual discrimination. You get into relationships with some of the other well-known types of things. Figure-ground can be a problem, which troubles me as a visual perceptual psychologist because figure-ground is a

119

pretty basic building block of all vision. If you have problems at that level, you can apply it onward and upward.

I am not sure that we are quite as good at this as we want to be. We use a variety of visual perceptual assessments. One of the things that we use is the Rivermead Post-Concussion Symptoms Questionnaire - and Morton Davis' Visual Scan Test. We do reading assessments using locally grown reading tests. One of the things worth noting here is by Susannah Trauzettel-Klosinski, who has come out with a nice photograph length reading test that is available in 17 different languages called IReST. It is one of the things that we are going to be incorporating into our work. Then you have speech and language evaluation for reading comprehension. We also use the NVT Vision Rehabilitation System for visual field defects and training.

I have to give a nod here to Ron Schuchard. The literature does not tell us really how to do rehabilitation for hemianopsia. Ron has put together a research project that is a randomized controlled study. Having complained about it, I now have to say, it is really needed because we are trying to provide the best patient care. The literature does not give us guidelines. I think this study will give you a scientifically sound basis for doing a better job in our rehabilitation program.

I also wanted to highlight that we start out with static assessments and then go to a variety of dynamic assessments - real world type tasks. This is, I think, an important part of a rehabilitation program. If you are a clinician and seeing people in your office, it is all well and good, and you can do good things; but quite often, there is a disconnect once that person walks out of your office. In our program we're trying to minimize that disconnect by taking what we have in the clinic and then moving it into the real world, looking for additional deficits that may not be obvious in the clinic setting but are obvious in a real world setting.

Optical Correction A Specific Example

Let me just skip through some of tasks that we are using so that I can look at one key study of a 28-year-old Iraq war veteran. He had right side frontal, parietal and occipital damage. Talking about TBI traumatic brain injury with a patient is much like telling somebody you have cancer. It all matters: They want to know where is it; what kind is it. With TBI, the patient may have a variety of "cancers," all at one time. This patient has a left homonymous hemianopsia and spastic hemiparesis. The best vergence at the acuity was 20/80. When we saw him on follow-up, his visual acuity actually improved. He had awareness and inattention problems and very severe PTSD. We did a variety of pre-trainings with him and assessments on the variety of tests. These two little video clips show the alphabet pencil test. What I want you to concentrate on is his struggle to go back and forth between the two pencils a few minutes into the pre-training exercise. Here he is a few minutes later - the timeframe here got compressed. We really worked well. But he is going back through with facility. You see a pause here and the pause comes from the fact that the letter pair happened to be 'FU'.

>> CHRISTOPHER TYLER:

Could you say once more what the task is?

>> GREGORY GOODRICH:

There are two pencils and alphabets that go down vertically. The task is to name the first letter on the first pencil and the first letter on second pencil. He goes through the entire pencil without any mistakes or any problems. So this does work. This is a one case example, but the whole process is one that I think needs to receive a lot more attention to put together something that we think is systematic. It covers a wide variety of present conditions. It addresses that immediate need of what do you when a patient walks in your door and says, "Hi, I'm here for rehabilitation".

Are we doing the optimum? Are we doing things that are effective? We try to develop an evidence base for everything that we are doing. On a clinical level, we have some evidence; but on a scientific level, we are very far that from that. Because a lot of the training we do is very time-intensive, one of the ways to look at the gaps is to take all of the information that has been presented here into the clinical settings so that we make sure all of the training that we are doing is appropriate, effective and optimum. It takes your patient's time, your time, and your resources.

If we're doing something in 12 weeks that we should be doing in four, then we need to know that as well. There is a great challenge in transferring all of this into the clinical rehabilitation setting.

>> JOHN BRABYN:

Thank you. We have time for one question.

Discussion

>> AUDIENCE MEMBER:

I am going to fit two questions in here. The first one is: Given all we heard about photophobia yesterday, is there anything other than tinted lenses that you are doing for that? Second, are any of these techniques really the same as or relatively minor modifications of the sorts of things that you do with the low vision patients?

>> GREGORY GOODRICH:

First question: we do not do a lot more than tints. I think one of the things worth mentioning is that you tell patients they can control lighting. You can do it with tints going indoor and outdoor. We'll do multiple tints. Suzanne talked about multiple tints yesterday. We have a different way of prescribing them, but it accomplishes the same thing.

>> AUDIENCE MEMBER:

Are the techniques really adaptations of what you have been doing for a long time, or are they in any way fundamentally different?

>> GREGORY GOODRICH:

From an instructor's point of view, one of the things that I like to communicate to our low vision therapists is that their concentration on the back of the eye, on what happened in an age-related disease, really should have been in the context of thinking about the brain and how people learn. With instructors, it is important to stress that it is all about the brain. In terms of rehabilitation, this is not about the eye.

There are some new things that we have brought in, mTBI being one of them. There are other things that are very similar to low vision, but the philosophy behind why you do it and how you do it changes. So, there is not a clear distinction in my mind. Some of it is new and some of it is old and adaptive. The adaptation is very critical.

>> AUDIENCE MEMBER:

You mentioned that mitosis may be taking place?

>> GLENN COCKERHAM:

We do see it with the Robo non-contact 9000 microscope. It is from a Japanese company. Suzanne has shared some papers from the '80s that were published in reputable journals that irrefutably showed mitosis. Are you really looking at the border at the endothelial cells? I would say it is putative. I see similar cells in some of these, having looked at probably 100,000 epithelial cells.

We also do manual counts and multiple tests for each eye looking for the test variables on the same day. All I know is that, instead of going down, the counts are going up. This is gratifying to me. They have all maintained their main point of transparency. So far, so good. We have not had to transplant any of them.

>> JOHN BRABYN:

Thank you very much. That was really a great talk. Our next speaker is Ken Ciuffreda who comes to us as a SUNY State College of Optometry in New York. When we were looking for someone to talk about the new and emerging techniques of oculomotor rehabilitation, oculomotor deficits, and their rehabilitation in mTBI, everyone pointed us in the same direction, and that was toward Ken. Thank you so much for coming.

Accommodative and Convergence Testing and Training in mTBI

KENNETH J. CIUFFREDA, OD, PhD

I would like to thank everyone at Smith-Kettlewell, my sort of father, that is binocular vision father, Dr. Jampolsky, and my binocular vision brother, Chris Tyler, who I have known for some time. During my time at Berkeley, we actually did some experiments with Alan Scott at SKIVS. We were looking at evidence for neurological control switching in saccadic dynamic overshoots. It feels like coming home to the Bay Area, Smith-Kettlewell, and Berkeley.

I would first like to look at the oculomotor deficits and give some examples of their remediation. Then I will talk about the therapeutic paradigm. Thankfully, Suzy's talk and Dr. Goodrich's talks segued nicely for me.

We have been doing this sort of testing and training for about 15 or so years. Most of what I will show is objective documentation: versional eye movements, vergence eye movements, and accommodation. In each case, I want to show one or two examples of the many examples that we have on the deficits that are found; and then remediation of the deficits, again looking at the clinical signs, symptoms, and concentrating on the objective measures; and then the overview of the training in the 10 microseconds of my talk!

Traumatic Brain Injury

I really did not want to get into TBI. I got forced into it by one of my clinical colleagues. I was more interested in basic vergence. This is one of the first patients that we saw, showing fixational movements on the midline pre-training. This gentleman was about 55-years old when he was attacked with a pipe on his head. He was a security guard in a housing center. You can see the improvement with training. This is the same patient with base-in prisms now added, showing a progressive improvement. This was done 15 or so years ago.

This is the pre and post horizontal eye position as a function of time, and vertical eye position as a function of time, in another TBI patient. It shows the improvement in fixation after about ten hours of vision training spaced out over an eight-week period, which is fairly evident. This is pre, and this is post. It does not get totally normal, but it does show marked improvement in the two-dimensional fixational abilities.

This is the Visagraph reading eye movement system. It shows a good example of reading in another TBI patient. This is at the end of the 10 weeks of training, about 10 hours total of training that was split between saccade training, fixation training, and simulated reading training. It looks more like a normal staircase now following the training than before. This particular subject had an improvement in reading of about 30-40 percent at the three-month follow up. All of the basic oculomotor parameter training changes persisted. The reading-related questionnaire showed marked improvement subjectively, and it too persisted.

Vergence

Turning now to vergence eye movements; we compared TBI and normal subjects. What is evident here, I think, is the time course of the vergence movement. The time course increased, and latency was increased, pre-training. That is, the overall response was slowed.

We are just about to embark on looking at training affects on vergence and accommodation. Our initial sample of 20 TBI subjects had both convergence and divergence deficits. Convergence and divergence were both slow. The results are published in JRRD. The remediation will be standard step tracking and prism flipper training, that we've all alluded to, in some of the other talks, and there are some really simple ways to get nice results objectively documented and consistent with what you find clinically with respect to signs and symptoms.

Accommodation

These are results we also published in JRRD. These are quasi-predictable steps. The stimulus timing was not predictable. We compressed time purposely to show the overall normal and abnormal profiles. It shows a square-wave improvement in a subject that did not show much of the deficit. There is some increased variability of accommodation and variability of vergence also. They get to their accommodation steady-state more slowly, and then they are unable to maintain it very well. This is one of the worst examples. There is a lot of variability at each of the end points. The trajectories are slower as well.

In contrast to normal subjects, those of a TBI individual show accommodation responses that are slow. It is reduced by about 50 percent in peak velocity. Time constants are also reduced by 50 percent. We were unable to get the latency here, but our prediction is that it would be increased like it was for vergence (about 100 milliseconds). Responses are slow, and there is more variability for accommodation than for vergence in these individuals.

We talked about some of this earlier. This table, compiled in a recent paper showing several studies, includes a section from the VA world. It is interesting that the estimates across each of the diagnostic areas are relatively similar. Whether it is a VA or non-VA population, blast or non-blast, we can probably use the same therapeutic interventions for the similar kinds of visual problems found.

Rehabilitation Studies

We also do vision rehabilitation. In one study, we had 33 subjects who finished OMT and having at least one sign or symptom improving markedly in terms of the clinical result and the patient's perspective. There was a 90% success rate. This kind of result is consistent across the board for this kind of therapy.

We do not have any data showing that we can remediate and dynamically accommodation, in TBI, although we do have data that we took over thirty years ago when I was at Berkeley. This is accommodation in patients with AI, but who do not have mTBI, with a step input from 1.5 to 4.5 diopters. Here the response is delayed. You can see it is slowed and variable, and then sustained reasonably well. Latency was also increased by 100 milliseconds. After about 10 hours of training, over a several week period of home-based only therapy, the responses normalized using objective evaluation. I think we can get the same result in soldiers with mTBI or other patients with neurological deficits. Of course, the results depend on the amount of time that we spend training them. Probably in our TBI cases, the impact might be a little less, and the time might take a little longer.

In terms of the training protocol, we have performed versional training for the past 10-15 years. It involves simple fixation of targets at different positions of gaze, and also predictable / non-predictable saccadic tracking horizontally and vertically. We also did pursuit, horizontal and vertically, at about 5 degrees per second; I do not think this is the most critical thing to do. It is, of course, important in terms of general oculomotor aspects. We also did simulated reading to get automaticity of eye movements.

Then we do vergence training. We do it therapeutically in terms of loose prisms, prism flippers, etc. We use step and ramp inputs. The step is important for initiation of the movement. The ramp is under feedback control for attaining bifixation within a few minutes of arc error at completion of the movement. We do something similar with step vergence. If I had one thing to pick, I would pick step, not ramp, training. It is more difficult, but more critical.

There is also vestibular-vergence interaction. As you get closer and closer with the target, the displacement between the center of rotation of the head and center of the eyes is no longer negligible. So we have to really to train VOR-vergence interaction at different distances, and also different directions of gaze.

Optical Correction

And in addition to the training, I give them a "crutch". We do not say that I want to give you a crutch because you will become dependent on it. No, we explain that a crutch initially helps to improve matters later on. We typically give plus lenses for near. Good numbers here are plus 1 and plus 1-1/4. We might also give prisms of 1-2 prism diopters in each eye.

For patients with mTBI and obviously those with stroke and hemianopia, where they have a shift of their egocentric space away from true zero, and it is shifted over to their seeing field, we use prisms to try and reduce the mismatch between the abnormal subjective and veridical objective visual space. This helps ambulation.

We also use tinted lenses for two areas. One is for patients with motion sensitivity problems. If we dim down the target, sometimes matters are improved. Tints are also used in cases of photosensitivity.

We also use field enhancing devices and sectoral occluders. A technique that we use is bi-nasal occluders for patients with visual motion sensitivity. When they work, they work really well. When they do not work, the person says, "I don't see anything different." In these latter cases, we move onto another dimension of the training or a device that we might help them during the remediation and afterwards.

Over the years, we have published several papers, and there are more to come that will show training protocols and positive training effects, as we are dealing with motor learning (Go to PUBMED). A paper that we published about 10 years ago is in Brain Research Protocols, and the current DoD funding for looking at vergence and accommodation remediation, as well as version and reading.

Discussion

>> *AUDIENCE MEMBER:*

When we evaluate these mTBI patients, we look at fixations qualitatively and watch the amplitude of their fixation. Do you have a quantitative way? What is your method for measuring and monitoring fixation ability?

>> *KENNETH CIUFFREDA:*

Clinically our quantitative options are not too good. Do we have increased drift or saccadic intrusion, or in that case nystagmus? Those are the three things we think about clinically. Saccadic intrusions are the same thing. I cannot suggest anything that is really good.

Some people have suggested just looking through a piece of plexiglass and watching it with targets on the plexiglass. Nothing more sophisticated than that. A ReadAlyzer is just an eye movement system that looks at eye system changes. No qualitative change, use the ReadAlyzer.

>> *AUDIENCE MEMBER:*

There is the potential to put everything on the computer instead of having a clinical evaluation system with a well-defined test. In your clinical judgment, how well are tests being performed. All of the indices are objective then.

>> *KENNETH CIUFFREDA:*

There are many pros that exist and Home Vision Therapy (HTS) makes it. You use targets that have good attention like a rapid serial visual processing (RSVP) kind of thing. We have computer control and we can make the target move at exactly 5 degrees per second. We can use the amplitudes we want and we try to use the object techniques concurrent with the clinical procedures. The pinnacle procedures are what everyone is going to look at when they text. The more standardization, and objective methods are going to be even better.

>> AUDIENCE MEMBER:

How did you measure divergence? The second one, accommodation relaxation was slow. Why is the relaxation slow when all of the stimuli and whatnot are not there? Are they all presbyopic?

>> KENNETH CIUFFREDA:

No. They were all optometry students, 20 years old. There is a lot of literature that would show relaxation of accommodation. People use the word "passivity," but it is not passivity. It is activity. Neurologically, it is not a passive movement. The time courses of divergence and convergence as well vergence are similar. Divergence is a tiny bit slower. It is not a passive movement.

>> AUDIENCE MEMBER:

How do you measure divergence?

>> KENNETH CIUFFREDA:

We use the masked eye movement systems that are around, as well as a device called the power refractor, which uses photo retinoscopy. Photo retinoscopy gives you pupil size, accommodation of state, refraction, or dynamically and horizontal vertical eye position.

>> JOHN BRABYN:

I think we need to stop here. Thank you very much.

Next we have Dr. Kim Cockerham, who is from the department of ophthalmology at Stanford. Kim is pretty well known in several areas of ophthalmology and everything that has to do with the orbit. She is going to fill us in on what can be done when the problems are the orbit affected by TBI.

Ophthalmic Interventions to Preserve Visual Function: Will Treating Fibrocyte-Mediated Inflammation Improve Visual Outcomes in Blast Injury and mTBI ?

KIMBERLY COCKERHAM, MD, FACS

Thank you very much. It is a real honor to be in this room with so many people that I have known for so many years and who are all interested in this tough topic.

Of course, I have to thank my husband for putting this meeting together. I actually was in the military for 15 years and I met Glenn there. I was involved in TBI when I was a fellow and a resident. At the time the U.S. was involved in Desert Storm One, Somalia and Bosnia. As a result, I have been involved in patient care and research that deals with TBI for quite a while.

The team that has contributed to the contents of this discussion includes Glenn and the team at the Palo Alto VA who are conducting a prospective research documenting the effects of blast injury and TBI on visual function of our returning Vets from Iraq and Afghanistan (a VA merit grant). A second funded project is the DoD project led by Ricardo Cavalarho, David Carpi and his team who have been developing eluting stents for ophthalmic use to prevent optic nerve damage and to minimize inflammation. Terry Smith and Ray Douglas have applied for DoD funding to study the role of fibrocyte-mediated inflammation and fibrosis.

Trauma-Based Fibrosis

Since I do not have a whole lot of time, this is going to be super brief. The bottom line is: how does trauma cause fibrosis? I want to define and describe the technology that has recently been funded by a DoD grant. My private practice is in Silicon Valley (and on the beach in Santa Cruz), so I am often surrounded by entrepreneurs who dream of improving the world with technology.

TBI is a common outcome of the IED's and other explosions. This signature wound of the current conflict is just tragic. We see these young adults who have their bodies protected by a Kevlar suit but their faces and limbs are exposed to blast injury. Too often they lose an eye, a portion of their face or their limbs that are less well protected. We do not have a lot of options to help these soldiers in the field minimize the impact of the blast to intact structures.

The currently funded Department of Defense grant is to use a rabbit model blast model and treat the eyes immediately with an episcleral eluting device to modulate inflammation and potentially minimize damage to the eyes. We are envisioning this being placed at the first battlefield station. This project requires a collaboration between brain injury thought leaders, basic scientists, clinicians and military leadership.

The mechanism of injury is blast damage resulting in inflammatory-mediated fibrosis and in some cases cell death. I am going to focus on the inflammatory-mediated fibrosis and using translation of eluting stents from other subspecialties. If you review the ophthalmic literature, there is increasing interest in this technology.

The role of fibrocytes and fibroblast has been characterized in burns, pulmonary disease and even thyroid disease. Graves' disease is often associated with a fibrocyte-mediated inflammation of the orbit (the area around and behind the eyeball).

Fibrosis in Graves' Disease

We know that the thyroid in Graves' Disease invites inflammation and fibrosis to occur around the eyes, the eyelids, the orbit and the pretibial area of the legs in particular. The fibrocyte used to be viewed as a particular thyrocyte that was very unique to thyroid inflammation. Now we know that trauma and other disease processes evoke fibroblasts from the bone marrow and home fibrocytes to the site of injury or inflammation. In the case of Graves' disease there are shared cell markers; trauma induced fibrocyte-mediated inflammation is less well understood.

If a patient with active thyroid disease (or autoimmunity due to an inflammatory or neoplastic process) undergoes surgery the inflammatory response is exaggerated. A blast injury from an IED is in an analogous fashion stimulate fibrocyte-mediated inflammation, fibrosis, and possibly apoptosis in the eye soft tissues, nerves and brain.

Where do the fibrocytes come from? At the time of trauma – or cell signaling in the case of autoimmune disease – the fibrocytes leave the bone marrow and enter the serum. They then travel to the site of trauma and cell signaling and induce inflammation, fibrosis and altered structure and function. If we could intercede quickly with an eluting device in the area that was traumatized, we could potentially prevent fibrocyte mediated disfigurement and disability.

Terry Smith, Ray Douglas and their group have studied these fibrocytes in Graves' disease for over a decade. They have more recently expanded their interest into how these "bad guys" mediate trauma related outcomes.

Fibrocyte quantification is done by taking peripheral blood specimens and then analyzing which subtypes are involved. In the initial studies in the group by Michigan, they traumatized the globe and found that CD45 is seen not only in thyroid related fibrotic problems, but also in trauma.

Episcleral Implants

IED related blast injuries result in varying degrees of TBI and ocular battlefield injuries of variable severity. Some ocular injuries are very obvious. Other patients' eyes appear normal but they are completely disabled by subtle double vision, nystagmus or optic nerve damage. The goal of the DoD episcleral implant is to create something that medics (or medical nurses or doctors without extensive ophthalmic training) could place right at the time of the injury - at the moment that they are first being evaluated in the field or just prior to getting their eye patch put on.

In ophthalmology, there have been eluting devices in the past. The Ocusert and Pilosert were available back when I was a resident in the 90s. They were used to combat dry eyes.

The goal of the current DoD project is to use a novel ocular drug delivery device that is field expedient. Potential small molecules for delivery include steroids, anti-fibrotics and unique estradiols. The first elutent will be brimonidine for neural protection.

We hope to establish safe and effective placement of the device in our New Zealand rabbit blast model and then pursue funding for human testing and eventually in country testing.

Discussion

>> AUDIENCE MEMBER:

I am not entirely sure how the direction you have taken with this project relates to the frequency with which the effects you are fighting really happen. Only 5 percent of TBI patients showed this.

>> KIMBERLY COCKERHAM:

The fibrocyte-mediated post-traumatic damage has at this point only studied in animal models. There has been some initial work done on non-blast injury. At this point, I cannot provide you with what percentage blast injured TBI patients will have occult eye injury that could be helped with this device. One hundred percent of patients with trauma will show increased fibrocytes. The level of the elevation in humans is at this point unknown - will it be a 5% or a 10% increase? We do not know.

>> AUDIENCE MEMBER:

There is a Walter Reed study that tries to correlate the incidence of ocular trauma concomitant with other injuries. If you have ocular trauma, you have a high incidence of traumatic brain injury, around 60-70% of the time. She was primarily looking at the ocular trauma component, but what about TBI and the fibrocyte population?

>> AUDIENCE MEMBER:

Kim, is there any literature on giving post-concussive patients a bolus?

>> KIMBERLY COCKERHAM:

It (providing high dose intravenous steroids for patients with suspected traumatic optic neuropathy) was done routinely. But this routine use of intravenous steroids did not demonstrate benefit when studied prospectively; also surgical decompression was also found to be of no benefit. But traumatic optic neuropathy is a real challenge to study because each injury is different. You cannot compare oranges to apples to bananas to pineapples.

>> AUDIENCE MEMBER:

Even in the theater, they backed off from the megadose steroids, even for spinal cord damage.

>> KIMBERLY COCKERHAM:

But they are still giving steroids – not in the very large "megadose" protocol but without good scientific evidence.

>> AUDIENCE MEMBER:

They have backed off and even changed the theater clinical practice guideline with the spinal cord protective megadose to prevent routine use of it even in spinal cord damage, where purportedly it was acceptable.

>> KIMBERLY COCKERHAM:

If tomorrow you were in a motorcycle accident and they take you to an ER with a concussion, do you want steroids? You are going to say yes. Not megadoses, but just routine dosing.

>> AUDIENCE MEMBER:

It makes a lot of sense to have someone that you can interview provide a contained level of the drug that is going to help you. It is obviously during that acute period. How do you define what is the duration is of acute period and can you then avoid further sequelae of the injury? Yesterday we were looking at biomarkers in the temporal way in relation to some of the phenomena that we are discussing. What biomarkers would you want to be looking at? You mentioned the marker on the fibroblasts for activation. What is the time course that you are really trying to intervene at? When do you know when it is over?

>> KIMBERLY COCKERHAM:

Looking specifically at biomarkers and the pilot studies, the window was about between 48-56 hours.

>> AUDIENCE MEMBER:

When does it end?

>> KIMBERLY COCKERHAM:

The effect trails off at about two weeks.

>> AUDIENCE MEMBER:

Your agent is basically neural protective. Does this mean there is no neural recovery mechanism?

>> KIMBERLY COCKERHAM:

The initial study is going to use brimonidine, fluorouracil, and dexamethasone. You can elute as many things as you elute. It is also possible to simultaneously elute antibiotic and anti-fungal drugs. But each drug will need to be tested independently.

>> AUDIENCE MEMBER:

Right. I did not catch the distinction between your mechanisms for neural protective versus neural recovery.

>> KIMBERLY COCKERHAM:

I am not the best one to detail Ricardo's work with this special estradiol but it acts at the level of the mitochondria to promote neural protection and recovery for intact but damaged nerves.

>> JOHN BRABYN:

Thank you very much Kim. That was great. Thank you. Next we have Gabrielle Saunders.

Comparison with Management of Hearing Impacts

GABRIELLE SAUNDERS, PhD

Auditory Rehabilitation for Blast-Exposed Veterans

As I mentioned earlier, in this presentation I am going to talk about auditory rehabilitation for blast-exposed Veterans who report hearing difficulties and yet have normal or almost normal hearing sensitivity. By 'almost normal' I mean that some individuals have slightly elevated thresholds above 4 kHz that has most likely been caused by exposure to noise. So, to remind you of the typical symptoms, these individuals report problems hearing in background noise, following rapid speech, following instructions, and following long conversations. They also report tinnitus and hyperacusis (hypersensitivity to loud sounds) – neither of which will I address today. Also as I said earlier, these difficulties, in the presence of normal hearing sensitivity tend to be indicative of auditory processing problems.

So, what is it we need to do with rehabilitation? For problems hearing in background noise ideally we would improve the signal-to-noise ratio (SNR), for difficulties following rapid speech we would improve temporal processing abilities and for problems following instructions and long conversations we would want to improve working memory.

What are the rehabilitation options with which these could be achieved? First, there is the use of FM (frequency modulation) technology to improve the SNR. FM technology uses FM waves to transmit signals picked up by a microphone worn by the talker, to a small receiver (that looks like a small hearing aid) worn by the listener. Unlike with a hearing aid, the signal is not amplified, but by having the microphone close to the talker's mouth and transmitting the signal via FM waves directly to the ear of the listener, the signal-to-noise ratio is dramatically improved. The advantage of FM technology is that, when used correctly it is highly effective; on the other hand it requires the user to carry with the assistive technology. Second, there is the option of using auditory training (AT), with which the user can train their listening and memory abilities. AT makes use of the fact that neurons in the brain can reorganize and restructure following intensive practice though strengthening of the neurological connections. Thus with an appropriate AT program it is possible that temporal processing and memory abilities could be improved. The potential advantage of AT is that if it is effective it is essentially a fix to the problem, the downside is that we don't know how effective it is, and that it takes extensive time (1 hr/day over 40 days) before any benefit is likely to be seen. However, for individuals with a recent traumatic brain injury it makes sense that the newly damaged neural pathways could be rebuilt using a program of intensive training. Finally, there is the possibility of using compensatory communication strategies, such as building vocabulary, actively listening, and problem solving, to improve communication skills.

So, the big question then is, are any or all of these interventions effective for blast-exposed individuals? I will now present some data from a study we have recently completed in which these interventions were examined, as follows. This work was conducted in a collaboration between NCRAR and the James A. Haley Veterans Hospital in Tampa, FL. We have data collected from 86 OEF/OIF Veterans, all of whom reported exposure to one or more blasts during active duty service, have normal hearing sensitivity but report auditory difficulties. Participants were randomly assigned to one of four intervention groups: (1) AT using the Brain Fitness Program from Posit Science and communication strategy training/informational counseling (AT-only), (2) FM system plus communication strategy training/informational counseling (FM-only), (3) FM system plus AT and communication strategy training/informational counseling (FM+AT), or (4) Communication strategy training/ informational counseling only (Controls). Participants underwent baseline testing, used the intervention to which they were assigned for 8-12 weeks, and then conducted outcomes testing. Tests included behavioral measures and self-report measures. It is interesting to note that we had a very difficult time recruiting individuals for this study, despite the fact that we hear from the audiology, TBI and social work clinics that they encounter many Veterans who meet our study criteria. We have concluded this is because these Veterans simply have so many stressors in their lives – PTSD, young families to care for, employment issues, and college attendance - that they have not got the time to participate.

First, did the participants use the intervention(s) they were assigned to? It turns out this is mixed. Less than 10% of individuals randomized to use AT (with or without FM) completed 75% or more of the recommended training, with over 60% completing less than 25% of the recommended training. The most common reasons given for not training were that the program was boring or difficult, or that there was not time to train. On the other hand, all but one individual assigned to use an FM system (with or without AT) actually used it. Most individuals (54%) used the system a few times a week, although some (15%) used it every day. The feedback we received about the FM system was in general very positive. And some individuals have gone on to acquire their own FM system since participating in the study.

In terms of benefit on the test measures, preliminary examination of the data shows that in general, the individuals in the FM+AT group had the best outcomes on behavioral and self-report measures, controls had the poorest outcomes, with those in the FM-alone and AT-alone groups falling between the two.

It seems then that the combination of AT and FM appears to be most effective form of intervention, but that there are considerable individual differences in compliance with the intervention and in outcome. It remains to be seen whether analyses examining compliance and outcome show a relationship between the two, and whether we can predict outcome from demographic factors or individual differences at baseline.

In summary, amplification is not the most appropriate form of intervention for blast-exposed veteran with normal hearing sensitivity and reports of hearing difficulties, but the combination of AT and FM appears to be effective. We need to update our current practices and then provide clinicians with evidence-based guidance so they know how best to provide audiologic rehabilitation to these patients. Thank you.

Discussion

>> AUDIENCE MEMBER:

I have a question about tinnitus. A lot of the patients have it and they are seeing audiology. They just say they learn to live with it. It is the same question that I have for the visual dysfunction. Do you have any idea what the natural history is?

>> GABRIELLE SAUNDERS:

I am not sure what you mean by natural history. If you mean, what is its etiology, that is complex and poorly understood. There are many factors associated with tinnitus but it is not clear how or where it is generated or how to stop it. It is likely there are multiple generators. In terms of whether it changes over time, again, there is no typical case and yes, most people simply have to learn to live with it.

>> AUDIENCE MEMBER:

But visual function certainly improves. I do not know why you would not get improvement in the 20-year-olds.

>> AUDIENCE MEMBER:

How closely are you working with the ENT and MD colleagues?

>> GABRIELLE SAUNDERS:

Right now we are working primarily with audiology.

>> AUDIENCE MEMBER:

In ophthalmology we have a classic disconnect that is very bad for TBI patients. Audiology and ENT seem to have a similar problem, at least in the VA. It would be nice if the kind of the team approach that I have started to do took hold and got us past this.

>> GABRIELLE SAUNDERS:

I would certainly agree. In fact I would suggest that teams should be broader still, and include other disciplines such as ophthalmology, neurology and psychiatry – depending of course on the patient's problems.

>> AUDIENCE MEMBER:

Do you have a systemic collection of information of all the VA's at this point?

>> GABRIELLE SAUNDERS:

The VA has an impressive electronic note system so in theory it is possible for a practitioner to examine the complete chart of any individual. In practice however, I do not think this is routinely done. Further, if IRB approvals are obtained it would also be possible to examine the many thousands of cases in VA databases.

>> GABRIELLE SAUNDERS:

This might really reflect my ignorance of the visual system. Is the training that you [Ciuffreda] described high-level training or more like the in-site training program that talks about visual attention to peripheral vision and responses?

>> KENNETH CIUFFREDA:

It mainly is the former, although attention is involved. The training has variations. You follow a target that is moving predictably stepwise on the screen either horizontally or vertically, or you follow a target that is moving smoothly 5-10 deg per second, or there is simulated reading where you are just seeing dot steps. Normal visual feedback is something we can see when they are following a finger. Then there is a crossover design to auditory feedback where we can effectively hear the eyes move. Some patients love hearing their eyes move. When we took away the auditory feedback, they asked how could they hear their oculomotor errors! You are always concurrently training general and visual attention.

>> GABRIELLE SAUNDERS:

Did their performance change when you removed the auditory signals?

>> KENNETH CIUFFREDA:

We looked at the first half only, when the auditory is added. It is done with any injury. Auditory enhanced results about 20 percent. This does not help as much I thought it would but it is a needed adjunct. The person gets it and likes it. He or she does not want to lose the additional feedback information. The improvements persist with feedback removed.

>> JOHN BRABYN:

Does anyone else have a question?

>> AUDIENCE MEMBER:

I feel terrible that, as an ophthalmologist, I am so ignorant about training. I certainly do not know auditory testing. I am upset about vision too. How do you really design an evidence-based testing program, one in which you can really tell the difference is because of a specific intervention rather than the placebo effect? How much is the Hawthorne effect? You said that since you have all of these different parameters of function that you look at you could, in fact, offer the person therapy for something that they do not have and see what it does to the things that they do have and vice versa. Has anyone used that kind of experimental design to comment on whether there would be a relevant way to address this problem versus the fact that you're sitting down with someone and attentive, giving close attention to them? Maybe it is the attention that is in fact making them better?

>> **KENNETH CIUFFREDA:**

One of these two approaches is where you have a crossover interventional design with a sham component. In this case, the subject is his or her own control. You could do a double baseline to make sure you have a good stable baseline. Then, adding objective measurements would clinch it. You do not worry about placebo effect because they cannot will the effect. They cannot "will" a slow saccade.

Panel Discussion

>> JOHN BRABYN:

Now we have Pia Hoenig on the panel from a popular vision clinic.

>> PIA HOENIG:

We have just begun, like many of you. We have known for a time that dynamics is one component of the training. We have shown in normals that even a normal healthy optometry student can improve the dynamics. What we are looking at now is non-traumatic brain injuries. Can they improve the dynamics?

We are comparing clinical training, the kind of crossover that Ken was talking about, baseline, independent, clinically, and in the lab. We then do the therapy independently. Then, they come back and we compare it again. As you know, the next step is going to be to try the find a group of TBI patients that have some commonalities. Finding commonalities is the hard part because it is such a multi-variable population. I will be happy to come back with that later on.

>> JOHN BRABYN:

Thank you. Did you have a question Ron?

>> RONALD SCHUCHARD:

The panel did an excellent job of answering a question related to this. We started talking about eye movements yesterday and today both Greg and Ken brought up eye movements. Ken, my concern is that you are recommending using the ReadAlyzer in the research labs. My understanding is that the ReadAlyzer really does not calculate all that much. It is looking at horizontal and I would say it is really for gross eye movements alone.

Would you like to comment in terms of the ReadAlyzer? I think it is a great clinical tool. I actually advocate it as a screening tool or as something to use in the clinic, but I am not so sure it is viable in the research lab.

I think it is a very nice tool to bring into the clinic because it does allow us to practice eye movements to some degree. We can look at patient's performance with it. We can gather some data out of that. There is some sense that what you really want in eye training targets the ReadAlyzer gives you. Maybe that is the kind of device we want, one that actually does tracking so I agree with you.

>> GLENN COCKERHAM:

I have to disagree with you. Not having vertical eye position is a limitation. But, besides being objective, the fact that the data automated in terms of the analysis with the reading grade depression, etc, is another nice thing about it. This gets rid of the experimental or clinical bias. I think the ReadAlyzer has more positives than negatives. Of course, I wish I had vertical eye position concurrent with the horizontal. But, in terms of gross eye movements, it records eye movements that are about half a degree or so in extent.

>> AUDIENCE MEMBER:

I would encourage you to double-check that.

>> GLENN COCKERHAM:

I've used it for 15-1/2 years.

>> AUDIENCE MEMBER:

I cannot get that.

>> GLENN COCKERHAM:

Do you use it in primary position or down gaze?

>> ARTHUR JAMPOLSKY:

Or reading.

>> GLENN COCKERHAM:

Do you do the reading in primary position or down gaze?

>> AUDIENCE MEMBER:

You do not stipulate to people how they read. We just ask them to read.

>> GLENN COCKERHAM:

One of the problems with any eye movement system is that when you are in downgaze but calibrated in primary position, you get other artifacts that make it very hard to come to a good conclusion.

>> AUDIENCE MEMBER:

This should not be the case in a good pupil eye chart.

>> GLENN COCKERHAM:

Well, your lid is going to cover as you go down the text. It is just mechanics.

>> JOHN BRABYN:

We can continue the advantages and disadvantages of eye tracking off line.

>> ARTHUR JAMPOLSKY:

When people read, they hold things in what is for them a reading position. I do not mean this facetiously. If you have large breasts, you hold it in a different place than if you are tall or something. If you are holding the reading, you are reading it differently than you do when you do not hold the reading. Reading is something that you have to replicate exactly as that individual reads. Some people read on a TV or a display system. The eyes are far from the reading display. To measure the final outcome in anything but the exact circumstance of the real world will leave you out in left field in the hurry. The same goes for the training. If you are going to train like the devil, fine. It seems logical to me that you would train in the exact real world environment too, if possible.

>> JOHN BRABYN:

Thanks. Vern?

>> VERNON ODOM:

Art, in some ways, this builds on your last comment. In terms of things like vision training, are there specific strategies and techniques that the person can use at home? Or, is everything done in the lab? Or, exactly what is the context of the training? For experimental purposes, one wants to do all of these very precise things.

>> KENNETH CIUFFREDA:

Well, we do our experiments only in the lab so that we can have control over everything that we do. We can time everything exactly: how many minutes of a fixation, saccade, training vergence, etc. One of the clinical problems is that in many cases you cannot do experiments unless you have to your patient come back to the clinic twice a week for an hour of training. In most cases, they will have a lot of home training. Life is more difficult if you do the home training because there is less control.

>> PIA HOENIG:

One additional component with the TBI patients is their difficulty in following directions about the order of things, as we discussed earlier. Where you are a healthy adult with the binocular vision problems we can send them home with quite complex instructions. Step by step reading instructions are a big problem with the TBI patient. It is not easy for them to figure out the directions.

>> GLENN COCKERHAM:

We also found that we get to a plateau level when we will send people home from Palo Alto to their home in Tennessee or . . .

>> KENNETH CIUFFREDA:

. . . Louisiana or wherever. So we are taking them out of their home environment, working with them, and sending them home again. When they come back 6-8 months later, their performance level is about where they were when they left us.

In our Binocular Vision Clinic, which is an outpatient program, getting people to come back is a problem. We do not know how to tell them if you do not come back every week, you are not going to get a benefit, but if you do come back every week, you are going to get a benefit. We are not sure that we can make that claim ethically. I think it is a real problem.

Studies that irrefutably show that what works and what does not would help. We also need studies that demonstrate how many times you need to see the person. There are some studies that are getting us towards this goal. There is a real need for the outpatient setting data.

>> SUZANNE WICKUM:

It can be hard to get these patients to do the therapy on their own. As Pia said, it is hard to get them to remember and understand the instructions. It helps if you have the ability to team with occupational or physical therapists. We have had a lot of experience working with the therapists. When the parents are going for their OT or PT sessions, the therapists can check in and reconfirm that the patients understand the therapy. They can even do some of the therapy with the vision. We have had good luck with the some the mild brain injury injuries.

>> GLENN COCKERHAM:

We know the medication affects the saccadic parameters. I would like the majority of the people that we see in the baseline exams we do see the effects of our anti-psychotics, and their phantom limb syndrome. A lot of Benadryl or some other drugs like that would stress the system. Is that a big issue as you go through this?

>> PIA HOENIG:

We all are nodding that it has a great effect. My own approach, and I think a lot of us would say this, is, well, we identified a problem. What is the cause, what is the lesion? That is what goes through our minds all of the time.

I would go for the more palliative approach - the prescription, the plus lens, the prisms - before we even start the therapy. I would like them to have less medication before I bring in the frontal lobe activities.

>> GLENN COCKERHAM:

You try to stop 48 hours before you do these tests?

>> PIA HOENIG:

If only I could wait until there is less medication on their scripts.

>> KIMBERLY COCKERHAM:

Are any of the teams using the quality-of-life tool in treatment guidance? It would be really helpful. Ophthalmologists thought that vision training does not work. This is because we see patients who say it does not work. We see patients who would be much happier monocular.

This is like an orthopedic surgeon saying, let's just cut off a limb. Some of these patients are tortured by the vision training. It is not helpful. They spend a lot of money and yet when you look at the quality-of-life tool, some of the happiest patients are the one-eyed patients.

In the eighties there was an occlusive contact lens. It sometimes gives you a very happy soldier. They are on medication and you can immediately have a patient be very, very happy with a single occlusive.

I am only certain that there is a lot of work being done with visual training in patients right now. The timing might not be right.

>> KENNETH CIUFFREDA:

I mentioned yesterday what happens if you occlude one eye. The answer is, if you have a vergence dysfunction, that is the deficit. Some of the young ones have a vergence deficit of saccade, pursuit, and fixational

146

ability so it probably will not help a whole lot of them. I teach normal and abnormal binocular vision. What I say is, if you have a patient who says, I really have to study for an exam, tell the patient to go and buy a one-dollar eye patch and put it on.

>> GABRIELLE SAUNDERS:

In response to your comment, it is the same in otology. Isn't the key for us to try and understand those individual differences so we do not throw everything at everybody?

>> GREGORY GOODRICH:

There is vision therapy and vision therapy and vision therapy. We do not have a consistent standard for what we mean by vision therapy. You have to wonder, well, if this patient was tortured by this therapist, and has since gone to someone else, would they have been happy with it in the end with the torturous therapist? We do not want to paint too broad of a brush and say everything does not work. The evidence base needs to be there. Certainly, if you're advocating that what you are doing is effective, you are obligated to provide the proof of that. Quite often in clinical practice you do something because you think it works. You do not always dot the I's and cross the T's to ensure that happens.

>> JOHN BRABYN:

It certainly means that there is a great need for clinical trial kind of thing or evidence-based research that will generate more evidence for treatments, rehab methods, populations and variations. Is this a fair comment? It seems like it applies to hearing as well as vision.

>> AUDIENCE MEMBER:

Actually, you raised a very good question. A lot of these patients are on a variety of drugs. We really have to do our homework to know at what the effects are of the commonly used medication. I am assuming that we are not going to talk about the illicit drugs, but only the medications that we are using now and how these may perturb a patient's auditory or vision function. Digoxin causes defects in vision and some other side effects. Have we really addressed the neuro, cognitive, visual processes side effects from the drugs that we are using to treat these soldiers?

>> KIMBERLY COCKERHAM:

That was a problem in TBI one, and it is a problem in TBI two. Everybody is gathering data in a different way. They do not know what medication the patients are on. If one thing comes out of this meeting I hope is that we work together and come up with a consensus. If we could develop a team approach for treatment and say, okay, what are the classic simple questions, and let's simply answer them. What are the effects of medication? It is not out there yet. What is the effect of vision training? Nobody says that this one causes accommodation and this has a pupillary effect. It is not out there yet. These are like the question of whether we should we use steroids in a post-traumatic case. All of these questions are basic questions. No one has sat back and taken the time to create a methodical assessment.

>> KENNETH CIUFFREDA:

That is a really good and difficult question. We tried to separate that out by noting the commonality of a drug and looking at therapeutic efficacy. The drugs our patients were taking was not the major factor causing the problem, but I say this tentatively.

Also, in our study on vergence and accommodation we know all of drugs that they are taking. Despite the constellation of all signs of drugs that they're taking, some of the findings converge for every single patient, those with hardly any drugs and those with the lot. Peak velocity or vergence or accommodation is object normal in every single patient of the 20-30 patients that we have done. You cannot have them come in with none of their drugs and ask similar questions with the therapy. Can you tell them not to do any of the therapy that they have been on prior to the 12 weeks?

>> AUDIENCE MEMBER:

I am not suggesting that we would not treat them. There is this weird effect of morphine in mice. Who would have expected that it would cause such light aversion? I would think that in some patients, anti-epileptic, antidepressant or all of these neurotropic or psychotropic drugs would unmask or exacerbate other underlying defects.

>> JOHN BRABYN:

For our next segment I would like to introduce a very wonderful person James Jorkasky. He is the executive director of The National Alliance for Eye and Vision Research (NAEVR). Jim and his organization have done a superb job of advocating for vision research as well as informing and educating people about the value of vision research of all kinds. This includes recent vision research on mTBI.

We are very lucky to have Jim here today. He was down in southern California yesterday receiving an award from the American Glaucoma Association for his work on behalf of vision research. He is the one guy who really has his finger on the pulse of funding from all sorts of research and alerting people if something is about to go wrong. I would also like to say that Jim was absolutely instrumental in putting this meeting together. He helped materially in not only telling us who we should get but how to get them.

>> JAMES JORKASKY:

I think he means you, Don [Gagliano].

>> JOHN BRABYN:

We owe a huge gratitude to you for that. We are really looking forward to your remarks on this topic.

Future Research Funding Possibilities in mTBI and Vision Function

JAMES JORKASKY

[Note: Due to numerous actions on this topic since the symposium was held, Mr. Jorkasky has abstracted the history of the VTRP from NAEVR's Value of Defense-Related Vision Research brochure, released in September 2012 and posted on its Web site at www.eyeresearch.org.]

What is the Vision Research Training Program (VTRP), and Why is it Important?

The dedicated VTRP budget line in Defense appropriations funds extramural vision research into immediate battlefield needs that is not conducted by the Department of Veterans Affairs (VA), elsewhere within the Department of Defense (DoD, including the Joint DoD/VA Vision Center of Excellence, VCE), the National Eye Institute (NEI) within the National Institutes of Health (NIH), or by private foundations. Although former Secretary of Defense Robert Gates identified Restoration of Sight and Eye-Care as one of four top priorities for deployment-related health research funding [with Traumatic Brain Injury (TBI), Post Traumatic Stress Disorder (PTSD), and Prosthetics], DoD has not yet established adequate "core" funding to address all vision research gaps, so VTRP funding is necessary.

The VTRP, which addresses deployment-related DoD-identified vision research gaps was established in Fiscal Year (FY) 2009 appropriations. Although the vision community has consistently requested $10 million in each funding cycle, annual appropriations have ranged from $3.25 million to $5 million. Vision, the sense most critical for optimal military performance in battlefield and support positions, is most vulnerable to acute and chronic injury. Research to effectively treat acute eye damage can have long term implications for an individual's vision health, productivity and quality of life for the reminder of military service and into civilian life.

Traumatic eye injury from penetrating wounds and TBI-related visual disorders ranks second only to hearing loss as the most common injury among "active" military:

- Traumatic eye injuries have accounted for upwards of 16 percent of all injuries in Operation Enduring Freedom (OEF) and Operation Iraqi Freedom (OIF).

- Male soldiers ages 20-24 account for 97 percent of visual injuries.

- Eye-injured soldiers have only a 20 percent return-to-duty rate as compared to an 80 percent rate for other battle trauma injuries.

- The VCE estimates that 58,000 enrolled OEF/OIF veterans have been diagnosed with eye conditions.

- The VA estimates that upwards of 75 percent of all TBI patients experience short- or long-term visual disorders including double vision, sensitivity to light, inability to read print, and other cognitive impairments.

What are the DoD-Identified Vision Research Gaps?

Ground soldiers especially face numerous assaults that potentially impair visual function, including:

- Eye injuries from chemical, biohazard, laser, and environmental exposure.

- Corneal (front-of-eye) and retinal (back-of-eye) injuries that are the result of direct blast injuries and are often not evaluated until a soldier's vital signs are first assessed and which, if not stabilized, lead to vision loss.

- Visual disorders as a result of Traumatic Brain Injury

- Potential long-term ocular injuries from a blast wave's pressure differential.

Due to the full spectrum of eye injuries–from superficial to blinding–as well as the military's desire to prevent injuries and to rehabilitate soldiers with injuries, the DoD has identified at least nine vision research gaps:

1. Inadequate mitigation and treatment of traumatic injuries, war-related injuries, and diseases to ocular structures and the visual system

2. Inadequate mitigation and treatment of visual dysfunction associated with TBI

3. Inadequate ocular and visual systems diagnostic capabilities and assessment strategies

4. Inadequate protection and prevention strategies

5. Inadequate vision rehabilitation strategies and quality of life measures

6. Lack of epidemiological studies of military eye trauma and TBI-related vision dysfunction

7. Inadequate vision restoration

8. Inadequate vision care education, training and simulation

9. Inadequate war fighter vision readiness and enhancement

How is the VTRP Managed?

The VTRP is managed by the DOD's Telemedicine and Advanced Technology Research Center (TATRC) within the U.S. Army Medical Research and Materiel Command (USAMRMC). TATRC, located at Fort Detrick, Maryland, added VTRP management to its existing Vision Research Portfolio (VRP), which had included other past Congressionally-directed program requests relating to vision research.

TATRC's VTRP Programmatic Committee, chaired by TATRC Director Colonel Karl Friedl, Ph.D. and Colonel Donald Gagliano, M.D., Director of the joint DOD/VA Vision Center of Excellence (VCE), consists of ophthalmic and optometric consultants to the Army, Navy and Air Force, as well as representatives from the NEI, the Food and Drug Administration (FDA) and stakeholders from the vision community [including the Association for Research in Vision and Ophthalmology (ARVO) and the National Alliance for Eye and Vision Research (NAEVR)]. The Committee develops a Program Announcement that seeks research proposals from vision researchers worldwide, evaluates the applicability of proposals to the DOD-identified vision research gaps, and determines awards after matching programmatic need with scientific peer review, which is conducted externally by the American Institute for Biological Sciences (AIBS).

Each year, TATRC representatives attend ARVO's annual meeting. In addition to speaking at a NAEVR-sponsored defense vision funding opportunities session, they also meet for more than 30 hours that week in one-on-one sessions with researchers to discuss the DOD gaps and research that may be responsive to those needs.

To Date, what has TATRC Awarded Vision Researchers from VTRP Funding?

In FY2009, with advocacy by NAEVR, Congress passed a Defense appropriations bill with the first-ever dedicated VTRP budget line, funded at $4 million. In FY2010, Congress funded the VTRP at $3.75 million. Each year, the VTRP was supported on a bipartisan basis in Congress, as well as by the Veterans Service Organizations (VSOs), including the Blinded Veterans Association (BVA).

In the FY2009-2010 VTRP funding cycle, TATRC announced twelve grants to vision researchers that totaled $11 million. This total reflected the FY2009 and 2010 Congressional VTRP appropriations of $4 million and $3.75 million, respectively, plus $4.1 million transferred over from TATRC's "sister" agency within the USAMRMC, the Clinical and Rehabilitative Medical Research Program (CRMRP), minus administrative costs. In adding funds to TATRC's portfolio, the CRMRP recognized the high quality of vision research grants (which scored at the highest percentile in peer review conducted by the AIBS) and their responsiveness to DoD research gaps. The twelve awardees reflect ophthalmic and optometric researchers from the US, Ireland, and Israel.

In FY2011, Congress funded the VTRP in Defense appropriations at $4 million and in FY2012 at $3.2 million. Once again, TATRC combined the FY2011 and 2012 appropriations and will award $14 million in grants to 21 vision researchers, reflecting the Congressional allocations plus $7 million transferred over from other DoD agencies, minus administrative costs. In this cycle, TATRC received more than 150 grant "pre-proposal" submissions in response to its Program Announcement, which specified two types of grants: Hypothesis Development Awards, funded up to $250,000 each, and Investigator-initiated Awards, funded up to $1 million each. TATRC plans to announce award recipients at the end of FY2012.

As of late 2012, DoD was operating on a FY2013 Continuing Resolution (CR) at the FY2012 funding level, and appropriations were not expected to be finalized until first-quarter of 2013. Earlier in the appropriations cycle, the House approved a Defense appropriations bill with VTRP funding at $10 million, while the Senate Appropriations Committee approved a bill that includes vision in a $50 million pool of funds for defense-related health research, not as a separate line item despite intense NAEVR advocacy.

Wrap-up Session

>> *WILLIAM GOOD:*

Thank you very much. Before we move on to our next section, Col. Gagliano has an announcement to make.

>> *DONALD GAGLIANO:*

I want to thank all the meeting organizers, participants and the one person who was the starting point, Dr. Art Jampolsky. Time and time again we heard about how influential he has been in so many careers of people in this room and also the careers of many other people that I've worked with who are not here. Dr. Art Jampolsky, can you come up here for a moment.

I mentioned on the first day that at that Art came up to the NAEVR booth at ARVO and said that I think I can post a West Coast meeting like the Schepens meeting. We challenged him to do it. I know he is going to say, I did it with the help of many others, but it does take a leader; you have been that for many people and for many as well. We have this tradition in the military of presenting coins to those who show exceptional performance. We have a Vision Centers for Excellence coin and very few of these are handed out outside of the organization. I think this time we have a person who really deserves one of these Vision Centers for Excellence coin.

The back is in Braille. This is the accepted logo from the Department of Defense and the Department of Veterans Affairs you've seen on several of the slides. The tradition is that we shake hands and I transfer the coin to you like that. [Applause] Those of you who have these coins need to remember that if we are out at dinner or a bar one day and somebody throws their coin on the bar, if you do not have yours, you buy.

>> *ARTHUR JAMPOLSKY:*

Thanks very much, Don. Now I have to declare a monetary interest in the meeting. Thank you very much.

>> *JOHN BRABYN:*

This would be a good opportunity to just say a few words of thanks to all the people involved in organizing this meeting. Art was the instigator and leader. He, of course, likes to push all the credit to other people, but he really deserves a huge amount of credit. Our Organizing Committee also included Bill Good and Christopher Tyler, not to mention Glenn Cockerham, Greg Goodrich, Ron Schuchard and Art Jampolsky. They helped us select all of the invitees, put the program together, and steered us in the direction of what topics we should be addressing, etc. The result was absolutely fantastic.

A number of Smith-Kettlewell staff were involved in the logistics of the meeting, led by Bebe St. John who spent a huge amount of time on it, Johnny, Dave, and Don with the audio visuals and technology; Doreen Polis and Connie Pierce are in the back doing the transcription, with Latricia Smith and Linda Washington. So there were a quite a number of people involved. I would like to give all of them special thanks.

We would like to wrap things up, so we will be out of here nearly on time.

We are fortunate in having with us today, as we did yesterday, Dr. Francis McVeigh. He is especially interested in keeping the list of gaps and unidentified problems that still need to have research done on them.

For this final session, Fran, would like to just give us a little summary of what they have come up with so far, what is their list of things that need to be done. And then if anyone from the audience has suggestions on things to add, we'll have an opportunity too for that.

>> FRANCIS McVEIGH:

I know you need time to ruminate on yesterday's conversations and today's new topics. Our gaps in the past were inadequate treatment of trauma injuries, war diseases related to the ocular structure the vision treatment, diagnostic treatment, and mitigation associated with TBI and war related injuries, inadequate vision restoration, population based studies, functional outcomes for ocular vision system, inadequate ocular diagnostics, and inadequate rehab strategies pathophysiology ocular injury. Also eye education care and training.

In our discussions, Dr. Kardon mentioned the need to look at the genetics aspects of this. Dr. Tyler mentioned functional imaging and the magnitude of the problem. Greg Goodrich and Dr. Good looked at biomarkers and asked how we rule things out and specificity. We also looked at the rehab approaches, efficacy and evidence-based topics. These were all sub categories of the main headings. At this time, I would like to ask Don and Rob if they have anything to add that you think we should look at. We can sit down as a group and prioritize these things with your help.

>> RANDY KARDON:

We should include quality of assessment tools, including multisensory quality-of-life assessment tools because it is hardly ever an individual dysfunction that we see in these patients. To my knowledge, nothing exists like that. We also talked about the correlation of ballistic eyeware in closed globe injury. There was no correlation. That, I think, derives a need for better pre-generalization strategies.

There is also the immune response to long-term affects on the immune system TBI. We just barely touched on the issues of neural fibular tangles and on the impact of multiple exposures and the development and redevelopment of neurofibrillary tangles, which, I think, have even greater opportunity for dysfunction. So I am going to put the immune response in there to follow the genetic response.

I would also add the photophobia response.

>> MICHAEL GORIN:

Diet-associated allodynia.

>> AUDIENCE MEMBER:

And the circadian disturbances.

>> FRANCIS McVEIGH:

A couple of allies were the polypharm aspects and how they relate to visual dysfunction. Lastly, the importance of multiple disciplines working as a team during treatment from the beginning to the assessments, the management, the treatment and rehab, the follow-up.

>> ROBERT MAZZOLI:

You covered it.

>> FRANCIS McVEIGH:

Did we miss anything?

>> AUDIENCE MEMBER:

I would like to make a case for the natural history of the vestibular effects, the objective findings. We mentioned specular microscopy. What happens to quality of life over time? A snapshot is all we have now. We have a small group, a small snapshot, but not the big picture. I think you need to know the natural history to know if the rehab is working or not.

>> *AUDIENCE MEMBER:*

There were some issues mentioned about sensory assessment when I was listening. People were talking about higher-level functioning being distressed in the auditory system. There are a whole series of tests out there used for treating visual information processing at various levels that might be useful in, first of all, identifying when those are problems and then monitoring treatment. Another issue is interaction of the various sensory systems. I remember there was some discussion of many vestibular functions and that some significant percentage of people had falls. A series of tests that either exist or could be created that would monitor this and relate what is happening to vision . . .

>> *KIMBERLY COCKERHAM:*

These are active duty soldiers. They can be told they need 12 vaccinations. I would like a baseline exam and I would store it on a card.

>> *GREGORY GOODRICH:*

And encrypted.

>> *KIMBERLY COCKERHAM:*

Could it be HIPAA compliant?

>> *GREGORY GOODRICH:*

I would like to see that baseline right after they walk out of a recruiter's office rather than at the end of basic training because it is training. It ought to be the recruiter's job to baseline all of this.

>> *KIMBERLY COCKERHAM:*

Yes. When they are recruited they will have to undergo an examination anyway. They are going to fill out this and have done that, so you can get the baseline at that point. I am only half-kidding. It is actually stored on a 'cloud'. We could all interface with this. I would love a cloud for the VA and the DoD research. We have this collection of people at each place and we do not know what happened at the prior place. It has been very disconnected.

>> *JOHN BRABYN:*

Thank you.

>> AUDIENCE MEMBER:

This may undermine what little credibility I may still have. The individuals who are serving in our military are the video game generation. There is good evidence to show that video games can alter tracking of objects and other things. I think we should be exploring how we can modify and utilize video games as a modality and a tool for functional assessment and rehabilitation. Unfortunately, the vision research community is not well connected to the video game development community, but the DoD could definitely remedy this.

>> ROBERT MAZZOLI:

I was just at Medicine Meets Virtual Reality (MMVR), which is the equivalent of the computer electronics show. It is a TATRC subsidized conference as well. At that meeting, a lot of the serious gaming developers are in the audience. One of the DoD attendees was actually using a video game called "Call to Duty" as an example. The anecdote was a 16-year-old kid who came up on a motorcycle accident and saw the victim bleeding out. The 16-year-old, who was never a Boy Scout and had no first aid training or any kind of formal or informal medical training, now had to take care of this patient. This victim was bleeding. The 16-year-old started to put pressure on the bleeding site. They asked him, "How did you know to do that?" Well, he explained that he played "Call to Duty" and one of the modules you have to take the medic training. In order to play the game, you have to do the medic training. This medic training that taught him how to do the pressure dressing was live in the video game and translated to the real world. They did not know that I was in the audience or that I am an ophthalmologist. After offering this example they then went on to say that ophthalmologists are pressing for an eye injury treatment where the idea is not to put a pressure dressing on, but to put a shield on. They challenged the video game developers to include what they said ophthalmologists are saying is the standard of practice. I thought that was great.

>> FRANCIS McVEIGH:

Russ Shelling, who is a navy commander at DARPA and has a video game background, is looking into different applications, not necessarily on vision, but across the board.

>> KIMBERLY COCKERHAM:

We presented there. There is a Kaiser Cadre who designed a baseball game. He was at the Academy of Ophthalmology and presented it to Gale Pollock.

>> FRANCIS McVEIGH:

The assistant surgeon general at one time?

>> KIMBERLY COCKERHAM:

So before the soldier went over, they did a baseline -- the saccades, pursuits, color, light. We were actually looking at it as a tool to see the role of the IED, which has not been addressed here. Your game score is too low, so there is a game out there. He was currently at USC. He is a fellow ophthalmologist.

>> AUDIENCE MEMBER:

First, I second or third your suggestions about the baseline. I think it is still a good idea. It should be practical. The sad fact is that the stumbling block is the word "agreement." We have to agree on what is a reasonable size for a test battery. If we could get ourselves together we would be able to do it.

Let me briefly address the game idea. I do not think it needs the tremendous sophistication that today's electronic computer games have. There has been attention to related research and it may well be able to serve a dual function: general permanence assessments and training. I have used such paradigms. It does not take a genius to have someone program this kind of stuff. For instance, interrelationship between cues is a possibility. Something that cues your attention to perceive something with better sensitivity than you would have without the cue within a few tenths of milliseconds. These are things that can be done and have been done.

If anybody is interested in these things and can get themselves moving in that regard, I could give them a lot of pointers. I know that there are other people who cold also.

>> KIMBERLY COCKERHAM:

Well, that is the upside of the baseball game. You can do visual field in patients, but if the patient is not attentive, the field is not useful. Because there is a probability that they get tired of watching things go up, I think the baseball thing is good. What would be good is to take that game and integrate all the things that everybody wants to have integrated into the game, but the guy just thinking he is whacking at the ball.

>> DEBORAH GILDEN:

Games like Wii and Xbox and so forth have been closed, but it was just announced that Microsoft is opening up the Xbox Kinect; so if you want to make your own or modify that, you can do it. The person giving the presentation said that they should not get these grids to do their range of exercises for cerebral palsy until they turned it into the game.

>> JOHN BRABYN:

Next time we have one of these meetings, we'll just turn it into the video.

On that note, I think we'd better let you go on time. I would like our co-chairman, Bill Good, to say a final word. Bill has been behind the scenes doing a lot work for this.

>> WILLIAM GOOD:

On behalf of the audience, the people who participated and those who came to listen to this conference, I want to say that this meeting really far exceeded any expectations that we had. It has really been a pleasure to meet the various people who work for the DoD, who are extremely high quality people. It is really clear that everyone is devoted to the right things, helping people who suffer from these trauma injuries. Thanks to you all for attending. We hope to see you again in probably two years.

If anyone is interested in taking a tour of Smith-Kettlewell, we will meet here at the front of the room in a couple minutes or so. We would be pleased to show you the various labs that we have and the kinds of things that we do here.

Acronyms

AI:	Accommodative insufficiency
APD:	Auditory processing disorders
ARVO:	Association for Research in Vision and Ophthalmology
AT:	Auditory training
BV:	Binocular vision
CDC:	Centers for Disease Control and Prevention
CDMRP:	Congressionally Directed Medical Research Program
CGRP:	Calcitonin gene-related peptide
CI:	Convergence insufficiency
CIT:	Critical Incident Technique
CN palsy:	Cranial nerve palsy
CSV:	(Vector Vision) Contrast Sensitivity Test
CSF:	Cerebrospinal fluid
DEM:	Development Eye Movement test
DoD:	Department of Defense
DVBIC:	Defense and Veterans Brain Injury Center
EMT:	Emergency Medical Technician
ENT:	Ear, Nose and Throat
ERG:	Electroretinography
FDT:	Frequency Doubling Technology
FIM:	Functional Independence Measure
FM:	Frequency modulation (FM) auditory systems
HIN:	Health Information Network
HIR:	Health Information Research
HTS:	Home Vision Therapy
IED:	Improvised explosive device
ipRGCs:	Intrinsically photosensitive retinal ganglion cells
IRB:	Institutional Review Board
LAA:	Light-associated allodynia
MEM:	MEM retinoscopy is a type of dynamic retinoscopy to assess lag of accommodation, in which the fixation target is a series of letters on the retinoscope or a card with letters at a normal reading distance.
MRMC:	(U.S. Army) Medical Research and Materiel Command (USAMRMC)

mTBI: Mild Traumatic Brain Injury

NAEVR: The National Alliance for Eye and Vision Research

NEI: National Eye Institute

NIH: National Institutes of Health

NOS: Neurological Outcome Scale

NOS-TBI: Neurological Outcome Scale for Traumatic Brain Injury

NVT: Neuro Vision Technology

OCT: Optical coherence tomography

OT: Occupational Therapy

PCR: Polymerase chain reaction

PERG: Pattern electroretinogram

PNS: Polytrauma Network Site

PT: Physical Therapy:

PTSD: Post-traumatic stress disorder

QUERI: Quality Enhancement Research Initiative. The QUERI centers focus on 9 high-risk and/or highly prevalent diseases or conditions among veterans

ROC: Receiver operating characteristic

RP: Retinitis pigmentosa

RSVP: Rapid serial visual processing

SF-36:   The SF-36 is a multi-purpose, short-form health survey with only 36 questions.

SKI: Smith-Kettlewell Institute

SNR: Signal-to-noise ratio

SSW: Staggered Spondaic Word test.

TATRC: Telemedicine & Advanced Technology Research Center. An office of the headquarters of the US Army Medical Research and Materiel Command (USAMRMC)

TBI: Traumatic brain injury

VA: Veterans Administration

VHA: Veterans Health Administration

References

Broglio SP, Macciocchi SN, Ferrara MS (2007) Sensitivity of the concussion assessment battery. Neurosurgery, 60:1050-1057.

Ciuffreda KJ, Han Y, Kapoor N, Ficarra AP (2006) Oculomotor rehabilitation for reading in acquired brain injury. NeuroRehab, 21:9-21.

Ciuffreda KJ, Rutner D, Kapoor N, Suchoff IB, Craig S, Han ME (2008) Vision therapy for oculomotor dysfunctions in acquired brain injury: a retrospective analysis. Optometry, 79:18-22.

Ciuffreda KJ, Yadav NK, Ludlam DP (2013) Effect of binasal occlusion (BNO) on the visual-evoked potential (VEP) in mild traumatic brain injury (mTBI). Brain Inj. 27:41-7.

Elbin RJ, Schatz P, Covassin T (2011) One-year test-retest reliability of the online version of ImPACT in high school athletes. Am J Sports Med, 39:2319-2324.

Fazio VC, Lovell MR, Pardini JE, Collins MW (2007) The relation between post concussion symptoms and neurocognitive performance in concussed athletes. NeuroRehab, 22: 207-216.

Galetta KM, Barrett J, Allen M, Madda F, Delicata D, Tennant AT, Branas CC, Maguire MG, Messner LV, Devick S, Galetta SL, Balcer LJ (2011) The King-Devick test as a determinant of head trauma and concussion in boxers and MMA fighters. Neurology. 76:1456-62.

Gallun FJ, Diedesch AC, Kubli LR, Walden TC, Folmer RL, Lewis MS, McDermott DJ, Fausti SA, Leek MR. (2012). Performance on tests of central auditory processing by individuals exposed to high-intensity blasts. J Rehab Res Devel, 49:1005-1024.

Gamlin PD (2002) Neural mechanisms for the control of vergence eye movements. Ann N Y Acad Sci 956:264-72.

Green W, Ciuffreda KJ, Thiagarajan P, Szymanowicz D, Ludlam DP, Kapoor N (2010) Accommodation in mild traumatic brain injury. J Rehab Res Devel, 47:183-99.

Han Y, Ciuffreda KJ, Kapoor N (2004) Reading-related oculomotor testing and training protocols for acquired brain injury in humans. Brain Res Brain Res Protocols, 14 :1-12.

Ho AM (2002) A simple conceptual model of primary pulmonary blast injury. Med Hypotheses, 59:611-3.

Katz J (1998) The SSW test manual. 5th ed. Vancouver (Canada): Precision Acoustics.

Killion MC, Niquette PA, Gudmundsen GI, Revit LJ, Banerjee S. (2004) Development of a quick speech-in-noise test for measuring signal-to-noise ratio loss in normal-hearing and hearing-impaired listeners. J Acoust Soc Am, 116:2395-405

Lew HL, Garvert DW, Pogoda TK, Hsu PT, Devine JM, White DK, Myers PJ, Goodrich GL (2009) Auditory and visual impairments in patients with blast-related traumatic brain injury: Effect of dual sensory impairment on the Functional Independence Measure. J Rehab Res Devel, 46:819-26.

Lew HL, Weihing J, Myers PJ, Pogoda TK, Goodrich GL (2010) Dual sensory impairment (DSI) in traumatic brain injury (TBI)--An emerging interdisciplinary challenge. NeuroRehab, 26:213-22.

Miller JR, Adamson GJ, Pink MM, Sweet JC (2007) Comparison of preseason, midseason, and postseason neurocognitive scores in uninjured collegiate football players. Am J Sports Med, 35:1284-8.

Musiek FE (1983) Assessment of central auditory dysfunction: the dichotic digit test revisited. Ear Hear, 4:79-83.

Musiek FE, Gollegly KM, Kibbe KS, Reeves AG (1989) Electrophysiologic and behavioral auditory findings in multiple sclerosis. Am J Otol, 10:343-50.

Musiek FE, Shinn JB, Jirsa R, Bamiou DE, Baran JA, Zaida E (2005) GIN (Gaps-In-Noise) test performance in subjects with confirmed central auditory nervous system involvement. Ear Hear, 26:608-18.

Owens BD, Kragh JF Jr, Wenke JC, Macaitis J, Wade CE, Holcomb JB. (2008). Combat wounds in operation Iraqi Freedom and operation Enduring Freedom. J Trauma, 64: 295-99

Sandel NK, Lovell MR, Kegel NE, Collins MW, Kontos AP (2011) The relationship of symptoms and neurocognitive performance to perceived recovery from sports-related concussion among adolescent athletes. Applied Neuropsychology: Child, 2012:1-6.

Saunders GH, Echt, KV. (2012). Blast exposure and dual sensory impairment: An evidence review and integrated rehabilitation approach. J Rehab Res Devel, 49:1043-1058.

Schatz P (2010) Long-term test-retest reliability of baseline cognitive assessments using ImPACT Am J Sports Med, 38:47-53.

Schatz P, Pardini JE, Lovell MR, Collins MW, Podell K (2006) Sensitivity and specificity of the ImPACT Test Battery for concussion in athletes. Arch Clin Neuropsychol, 21:91-99.

Taber KH, Warden DL, Hurley RA. (2006). Blast-related traumatic brain injury: what is known? J Neuropsychiat Clin Neurosci, 18:141-45.

Thiagarajan P, Ciuffreda KJ, Ludlam DP (2011) Vergence dysfunction in mild traumatic brain injury (mTBI): a review. Ophthal Physiol Opt, 31:456-68.

Index